02-CTS-4a

MEMBRANE BIOREACTORS FOR ANAEROBIC TREATMENT OF WASTEWATERS: PHASE II

Eric R. Hall
University of British Columbia

Pierre R. Bérubé
University of British Columbia

2006

WERF

The Water Environment Research Foundation, a not-for-profit organization, funds and manages water quality research for its subscribers through a diverse public-private partnership between municipal utilities, corporations, academia, industry, and the federal government. WERF subscribers include municipal and regional water and wastewater utilities, industrial corporations, environmental engineering firms, and others that share a commitment to cost-effective water quality solutions. WERF is dedicated to advancing science and technology addressing water quality issues as they impact water resources, the atmosphere, the lands, and quality of life.

For more information, contact:
Water Environment Research Foundation
635 Slaters Lane, Suite 300
Alexandria, VA 22314-1177
Tel: (703) 684-2470 Fax: (703) 299-0742
www.werf.org werf@werf.org

This report was co-published by the following organization. For nonsubscriber sales information, contact:

IWA Publishing
Alliance House, 12 Caxton Street
London SW1H 0QS, United Kingdom
Tel: +44 (0) 20 7654 5500
Fax: +44 (0) 20 7654 5555
www.iwapublishing.com
publications@iwap.co.uk

Library of Congress Catalog Card Number: 2006924065
IWAP ISBN: 978-1-84339-760-1

The University of British Columbia

The research on which this report is based was funded, in part, by the United States Environmental Protection Agency through Cooperative Agreement No. CR-827345-01 with the Water Environment Research Foundation (WERF). Unless an EPA logo appears on the cover, this report is a publication of WERF, not EPA. Funds awarded under the Cooperative Agreement cited above were not used for editorial services, reproduction, printing, or distribution.

ACKNOWLEDGMENTS

This research was funded by the Water Environment Research Foundation

Report Preparation

Principal Investigator:
Eric R. Hall
University of British Columbia

Co-Investigators:
Pierre R. Bérubé
University of British Columbia

Project Subcommittee

Robert Reich, P.E. (Research Council Liaison)
DuPont Company

Julian Sandino, P.E., Chair
CH2M Hill

Dick Speece, Ph.D.
Vanderbilt University

Heath Mash, Ph.D.
U.S. EPA

Manuel S. Rodriguez Susa, Ph.D.
Universidad de los Andes — Bogota, Colombia

Paul Sun, P.E., Ph.D.
Shell Global Solutions, Inc.

Theodore Papagiannis, P.E.
Sanitation Districts of Los Angeles County

The project team would like to thank US Filter Memcor, Ionics/Mitsubishi, Sanitherm Engineering Ltd. (i.e., Toray membranes), ITT Industries/PCI Membrane Systems, Inc., and Koch Membrane Systems Inc. for providing the bench-scale membrane modules used in the present study.

Water Environment Research Foundation Staff

Director of Research: Daniel M. Woltering, Ph.D.
Program Managers: Mary Strawn, Christine Handog

ABSTRACT AND BENEFITS

Abstract:

The specific objectives of the present study were 1) to assess and compare the treatment performance of external and submerged membrane anaerobic membrane bioreactors (AnMBRs) operated at different organic loading rates (OLRs) when treating a low-strength wastewater at a relatively low temperature, 2) to assess and compare the membrane filtration characteristics of mixed liquors generated in external and submerged membrane AnMBRs, 3) to assess and compare the membrane filtration characteristics of a mixed liquor in AnMBRs when filtering through inorganic and organic membranes, and 4) to assess and compare the membrane filtration characteristics of the mixed liquors generated in AnMBRs to the mixed liquor generated in an aerobic membrane bioreactor (MBR) operated with the same influent wastewater.

The study results indicated that treated effluent quality did not appear to be significantly affected by the type of membrane chosen for the application; rather, effluent quality was most impacted by the operation and performance of the anaerobic bioreactor component of the system. The data available indicated that, on average, AnMBR treatment resulted in 72% removal of the raw wastewater chemical oxygen demand (COD), leaving a residual COD concentration of about 73 mg/L in the permeate. The limited amount of 5-day biochemical oxygen demand (BOD_5) data indicated that an AnMBR could achieve about 65% BOD_5 removal, leaving a permeate BOD_5 concentration that averaged 56 mg/L. The estimated BOD_5/COD ratio of 0.76 for the AnMBR permeate indicated that much of the organic matter that was refractory to anaerobic treatment, was in fact, aerobically biodegradable. This suggests that aerobic post-treatment may be needed to achieve a secondary quality effluent in terms of BOD_5 or COD. The operating conditions and anaerobic treatment performance achieved in the study were comparable to those reported in the very small number of similar previous studies that were cited in the analysis of the relevant compiled literature.

Membrane flux became the limiting process attribute. The permeate flux that could be maintained in the external membrane AnMBR was higher than that which could be maintained in the submerged membrane system. It should be noted that the external membrane AnMBR was operated at a much higher trans-membrane pressure (TMP) than the submerged system. Also, the membrane flux maintenance appeared to be superior in the ceramic membrane relative to that in the polymeric membrane. However, based on discussions with the Project Subcommittee and a review of relevant literature, it was concluded that the high capital and operating cost associated with external inorganic membranes outweighed the lower fouling tendency of these systems. The measured resistance of anaerobic mixed liquor to membrane filtration was much greater than that of mixed liquor from an aerobic MBR treating the same wastewater, and as a result, the AnMBR fouled more readily than an aerobic MBR. The maximum permeate flux that could be sustained in a submerged membrane AnMBR was approximately 5 $L/m^2 \cdot hr$, while the permeate flux that could be sustained in a submerged membrane aerobic MBR was at least 15 $L/m^2 \cdot hr$. Although the concentrations of soluble proteins in the mixed liquor for the AnMBR were only slightly higher than those for the aerobic MBR, these proteins accumulated to a greater extent on the membranes in an AnMBR than on the membranes in an aerobic MBR. However, the overall concentrations of total organic carbon (TOC) measured in the foulant material extracted from the membranes for both the AnMBR and the aerobic MBR were relatively similar. The results suggested that inorganic foulants (e.g., struvite) were responsible for the greater fouling propensity of an AnMBR when compared to that of an aerobic MBR, that that removal of struvite governed the recovery of permeate flux in an AnMBR and that struvite was predominantly associated with internal membrane fouling. Eventually, irreversible fouling was observed for all of the membranes used in the present study.

Benefits:

- Demonstrates that the treatment performance achieved in the present study did not materially exceed that of the few related studies reported in the literature and, therefore, this study does not point to the possibility of achieving secondary quality effluent in an economical manner using AnMBR technology alone.
- Demonstrates that there is likely merit in assessing the efficacy of an anaerobic system, such as an anaerobic moving bed biofilm reactor to remove the bulk of the organic material from a wastewater, coupled to an aerobic MBR to polish the effluent from the anaerobic bioreactor.

Keywords: Anaerobic, filtration characteristics, low-strength wastewater, membrane bioreactor, treatment performance.

TABLE OF CONTENTS

LIST OF TABLES

LIST OF FIGURES

LIST OF ACRONYMS

AnMBR	Anaerobic membrane bioreactor
BOD	Biochemical oxygen demand
BOD_5	5-day biochemical oxygen demand
CFV	Cross-flow velocity
COD	Chemical oxygen demand
CBOD	Carbonaceous biochemical oxygen demand
$CBOD_5$	Carbonaceous 5-day biochemical oxygen demand
DDW	Distilled de-ionized water
HRT	Hydraulic retention time
J	Permeate flux
MBR	Membrane bioreactor
MLSS	Mixed liquor suspended solids
OLR	Organic loading rate
PSC	WERF Project Subcommittee
PVDF	Polyvinylidene difluoride
Ra	Resistance due to adsorption
Rc	Resistance due to cake layer
Rm	Resistance due to membrane
Rp	Resistance due to pore plugging
Rs	Resistance due to foulants
SEM	Scanning electron microscope
EDX	Energy dispersive x-ray
SMP	Soluble microbial product
SRT	Solids retention time
SS	Suspended solids
TKN	Total Kjeldahl nitrogen
TMP	Trans-membrane pressure
TOC	Total organic carbon
TP	Total phosphorus
TSS	Total suspended solids
UBC	University of British Columbia
UASB	Upflow anaerobic sludge blanket
VFA	Volatile fatty acids
VSS	Volatile suspended solids

EXECUTIVE SUMMARY

Anaerobic biological treatment systems can offer a number of economical and operational advantages over their aerobic counterparts. However, to date, the use of conventional anaerobic biological systems for the treatment of dilute wastewaters has been relatively limited, especially in cooler climates. This is largely because under such conditions, conventional anaerobic biological treatment systems cannot sustain sufficiently large biomass concentrations to effectively treat the wastewater. On the other hand, in a membrane bioreactor (MBR) MBR, the membrane component of the system can retain virtually all of the biomass within the bioreactor. Therefore, it should be possible to maintain an adequate biomass concentration by coupling a membrane to an anaerobic biological reactor to treat low-strength wastewaters effectively in cooler climates. In addition, it may be possible to overcome some of the treatment limitations associated with anaerobic systems by coupling a membrane to the bioreactor. Anaerobic biological treatment systems can effectively remove the bulk of the organic contaminants present in a wastewater; however, they typically are not effective at removing soluble and colloidal organic contaminants to the same levels that are achievable with aerobic biotreatment systems. The results from recent research suggest that by coupling a membrane with an anaerobic bioreactor, residual organics could be retained in the system independently of the hydraulic throughput, enabling these contaminants to be hydrolyzed and biodegraded. Permeate flux maintenance is one of the most significant factors that affects capital and operating costs associated with an MBR. Accordingly, extensive research has been completed to investigate parameters that affect permeate flux in aerobic MBRs. In contrast, only a limited amount of research has focused on parameters that govern permeate flux in an AnMBR.

An analysis of the relevant compiled literature and other information, on the topic of anaerobic MBR wastewater treatment, was completed and published as a separate WERF Report (WERF Project 02-CTS-4 Task 1. *Phase 1 Report: Compilation/Review of Existing Literature*). The principal objective of the information analysis was to identify the variables that appeared to be critical to achieving a high effluent quality in an economical fashion, in applying the AnMBR technology for treatment of low strength wastewaters. The results from the information analysis clearly implied that AnMBR technology has the potential to achieve near complete treatment (i.e., $CBOD_5$ less than or equal to 30 mg/L) of municipal and/or low-strength industrial wastewaters operating under relatively low temperature conditions (25–35°C). The results also indicated that that the efficiency of the membrane component likely will dictate treatment system economics.

The overall objective of the present research was to address the AnMBR technology issues and research needs identified during the literature review. The specific objectives were to 1) to assess and compare the treatment performance of external and submerged membrane AnMBRs operated at different organic loading rates (OLRs) when treating a low strength municipal wastewater at a relatively low temperature, 2) to assess and compare the membrane filtration characteristics of mixed liquors generated in external and submerged membrane AnMBRs, 3) to assess and compare the membrane filtration characteristics of a mixed liquor in AnMBRs when filtering through inorganic and organic membranes, and 4) to assess and compare the membrane filtration characteristics of the mixed liquors generated in AnMBRs to the mixed liquor generated in an aerobic MBR operated with the same influent wastewater.

The results indicated that the treated effluent quality did not appear to be significantly affected by the type of membrane chosen for an AnMBR application; rather, effluent quality was most impacted by the operation and performance of the anaerobic bioreactor component of the system. The potential for COD removal in an AnMBR could be judged best from the data collected during the period of stable performance that was associated with Phase II of the study. When the influent was supplemented with acetate, the average total COD removal achieved in the submerged membrane AnMBR was about 80%. Measurements of the COD concentration of the raw wastewater permitted the estimation of the removal efficiency for COD that originated from the wastewater itself. The data available indicated that, on average, AnMBR treatment resulted in 72% removal of the raw wastewater COD, leaving a residual COD concentration of about 73 mg/L in the permeate. A limited amount of 5-day carbonaceous biochemical oxygen demand ($CBOD_5$) data indicated that a submerged membrane AnMBR could achieve about 65% $CBOD_5$ removal, leaving a permeate $CBOD_5$ concentration that averaged 56 mg/L. The estimated $CBOD_5$/COD ratio of 0.76 for the AnMBR permeate indicated that much of the organic matter that was refractory to anaerobic treatment, was in fact, aerobically biodegradable. This suggests that aerobic post-treatment may be needed to achieve a secondary quality effluent in terms of $CBOD_5$ or COD. The operating conditions and anaerobic treatment efficiencies achieved in the study were comparable to those reported in the very small number of similar previous studies that were cited in the analysis of the relevant compiled literature (WERF Project 02-CTS-4 Task 1. *Phase I Report: Compilation/Review of Existing Literature*). Measured biogas production rates were always lower, and in some cases substantially lower, than the expected values. This was attributed to the fact that when low-strength wastewaters are treated anaerobically, a substantial portion of the CO_2 and the CH_4 generated will exit the process dissolved in the treated effluent. In addition, it was recognized that some of the retained particulate COD that was included in the removal efficiency calculation may not be biodegradable and fermentable under the conditions applied.

Membrane flux became the limiting process attribute for the AnMBR. The permeate flux that could be maintained in the external membrane AnMBR (50 to 100 $L/m^2 \cdot hr$) was higher than that which could be maintained in the submerged membrane system (5 $L/m^2 \cdot hr$). It should be noted that the external membrane AnMBR was operated at a much higher TMP of (40 psi or 275 kPa) than the submerged system (1 psi or 6.9 kPa). Also, membrane flux maintenance appeared to be superior in the ceramic membrane relative to that in the polymeric membrane in the external membrane AnMBR. The permeate flux that could be maintained for the polymeric membranes was approximately 50 $L/m^2 \cdot hr$, while that which could be maintained for the ceramic membranes was approximately 100 $L/m^2 \cdot hr$ (at a TMP of 40 psi or 275 kPa). However, based on discussions with the Project Subcommittee (PSC) and a review of relevant literature, it was concluded that the high capital and operating cost associated with external inorganic membranes outweighed the lower fouling tendency of these systems. Irreversible fouling was the major contributor to flux decline in both the external and submerged membrane AnMBR systems and this fouling could not be controlled sustainably by chemical cleaning. Eventually, irreversible fouling was observed for all of the membranes used in the present study.

The submerged membrane AnMBR fouled more readily than the parallel submerged membrane aerobic MBR. The maximum permeate flux that could be sustained in an AnMBR was approximately 5 $L/m^2 \cdot hr$, while the permeate flux that could be sustained in the aerobic MBR was at least 15 $L/m^2 \cdot hr$ (at a TMP of approximately 1 psi or 6.9 kPa). Based on discussions with the supplier of the submerged membranes used during Phase II of the study, it was recognized that an operating permeate flux in the order of 15 to 30 $L/m^2 \cdot hr$ can typically be maintained for full-scale sparged submerged hollow fiber aerobic MBRs treating municipal wastewaters. Although the concentrations of soluble proteins in the mixed liquor from the submerged membrane AnMBR were only slightly higher than those from the parallel aerobic MBR, these proteins accumulated more on the membranes in the AnMBR than on the membranes in the aerobic MBR. However, the overall concentrations of TOC measured in the foulant material extracted from the membranes from both the AnMBR and the aerobic MBR were similar, indicating that the overall amounts of organic foulants that accumulated on the membranes for both systems were similar. Inorganic foulants (e.g., struvite) may be at least partly responsible for the greater fouling propensity of an AnMBR when compared to that of an aerobic MBR. The results suggest that removal of struvite during membrane cleaning governs the recovery of permeate flux. Struvite (i.e., phosphorus) was predominantly associated with internal membrane fouling.

CHAPTER 1.0

INTRODUCTION

Anaerobic biological treatment systems can offer a number of advantages over their aerobic counterparts. The operational costs associated with anaerobic systems are typically lower than those of aerobic systems, and anaerobic systems also generate less waste sludge. In addition, the energy associated with the biogas produced during anaerobic biological treatment can potentially be recovered. However, to date, the use of conventional anaerobic biological systems for the treatment of dilute wastewaters has been relatively limited. This is particularly true for the treatment of low-strength wastewaters in cooler climates. Under such conditions, the biomass growth yield and growth rate are relatively low, resulting in low overall net biomass production. The net biomass production must exceed the net biomass loss to the effluent for a biological treatment system to function properly. However, in conventional anaerobic biological treatment systems, the net biomass loss to the effluent often is governed by the relatively poor settling characteristics of the biosolids. As a result, it is typically not possible to maintain a sufficiently large biomass concentration in a conventional anaerobic biological system to treat low-strength wastewaters effectively in cooler climates. On the other hand, in an MBR, the membrane component of the system can retain virtually all of the biomass within the bioreactor. Therefore, it should be possible to maintain an adequate biomass concentration by coupling a membrane to an anaerobic biological reactor to treat effectively low-strength wastewaters in cooler climates. In addition, it may be possible to overcome some of the treatment limitations associated with anaerobic systems by coupling a membrane to the bioreactor. Anaerobic biological treatment systems can effectively remove the bulk of the organic contaminants present in a wastewater; however, they are typically not effective for removing soluble and colloidal organic contaminants to the same levels that are achievable with aerobic biotreatment systems. The results from recent research suggest that by coupling a membrane with an anaerobic bioreactor, residual organics could be retained in the system independently of the hydraulic throughput, enabling these contaminants to be hydrolyzed and biodegraded (Stuckey and Hu, 2003).

Permeate flux maintenance is one of the most significant factors that affects capital and operating costs associated with an MBR. Accordingly, extensive research has been completed to investigate parameters that affect permeate flux in aerobic MBRs. Much of the recent research of this type is summarized in Chang et al. (2002). In contrast, only a limited amount of research has focused on parameters that govern permeate flux in an AnMBR.

1.1 AnMBR Technology Issues and Research Needs

An analysis of the relevant compiled literature and other information, on the topic of anaerobic MBR wastewater treatment, was completed and published as a separate WERF Report (WERF Project 02-CTS-4 Task 1. *Phase 1 Report: Compilation/Review of Existing Literature*). The principal objective of the information analysis was to identify the variables that appeared to be critical to achieving a high effluent quality in an economical fashion, in applying the AnMBR technology for treatment of low strength wastewaters. The results from the information analysis implied that AnMBR technology has the potential to achieve near complete treatment (i.e., removal of $CBOD_5$ to less than 30 mg/L) of municipal and/or low strength industrial wastewaters operating under relatively low temperature conditions (25–35°C). The results also indicated that the efficiency of the membrane component likely would dictate treatment system economics.

Based on the information analysis, the parameters that were identified as being important for the performance of the bioreactor and/or the membrane component of an AnMBR are listed below.

1.1.1 Critical Bioreactor Parameters

- **Bioreactor Operating Temperature.** The information analysis clearly indicated that bioreactor operating temperature will significantly affect treatment efficiency. The results imply that AnMBR technology should be capable of achieving near complete treatment of low strength wastewaters operating under relatively low temperature conditions. However, the level of treatment that can be achieved in an AnMBR treating low strength wastewaters operating under relatively low temperature conditions needs to be established.
- **Bioreactor OLR.** The bioreactor COD, OLR, and the operating solids retention time (SRT) dictate treatment performance. Results from the information analysis suggest that the anaerobic bioreactor component of an AnMBR system should be capable of operating up to approximately 5 kg $COD/m^3 \cdot day$ and achieve near complete treatment of municipal wastewater under relatively low temperature conditions. The effect of the OLR on the level of treatment that can be achieved in an AnMBR treating low strength wastewaters operating under relatively low temperature conditions needs to be established.
- **Bioreactor Configuration and Biomass Characteristics.** The membrane component plays a critical role in dictating the performance and efficiency of an AnMBR, allowing for capture and hydrolysis of wastewater organic colloids and particulates, physical retention of certain soluble organics and maintenance of a high concentration of active methanogens in the bioreactor component. However, the inclusion of a membrane component may also reduce biomass activity due to the resulting high shear environment. The effect of different membrane module configurations on the level of treatment that can be achieved in an AnMBR treating low-strength wastewaters operating under low-temperature conditions needs to be established.

1.1.2 Critical Membrane System Parameters

- **Membrane Module Configuration.** Both external membrane systems and submerged membrane systems impart a significant amount of shear force on the biomass in an AnMBR. High shear forces can reduce the microbial activity in an AnMBR, reduce the size of the biosolids floc in the mixed liquor and increase the release of soluble microbial products, all of which have been documented to affect permeate flux. A comprehensive comparison of the effect of different membrane configurations on the permeate flux in an AnMBR treating low strength wastewaters operating under relatively low-temperature conditions needs to be performed.

- **Membrane Material.** The type of membrane material used can significantly affect fouling in an AnMBR. Fouling has been documented to be lower for inorganic membranes. However, the capital costs associated with inorganic membranes are substantially higher than those for polymeric membranes. The effect of different membrane materials on the permeate flux in an AnMBR treating low strength wastewaters operating under relatively low temperature conditions needs to be established.

1.2 Objectives

The overall objective of the present research was to address the AnMBR technology issues and research needs identified during the literature review. The specific objectives were to assess and compare:

- treatment performance of external and submerged membrane AnMBRs operated at different OLRs when treating a low strength municipal wastewater at a relatively low temperature;
- membrane filtration characteristics of mixed liquors generated in external and submerged membrane AnMBRs;
- membrane filtration characteristics of a mixed liquor in AnMBRs when filtering through inorganic and organic membranes; and
- membrane filtration characteristics of the mixed liquors generated in AnMBRs to the mixed liquor generated in an aerobic MBR operated with the same influent wastewater.

An experimental program was developed to address the objectives listed above. A description of the experimental program is presented in Section 2.1. One of the main outcomes of this experimental program, hereafter referred to as Phase I of the study, was the realization that the operation of an AnMBR is highly impacted by the permeate flux that can be maintained in the system. To further characterize the membrane filtration characteristics in an AnMBR treating low strength wastewaters operating under relatively low temperature conditions, a follow-up experimental program, hereafter refered to as Phase II of the study, was completed. A description of the experimental program for Phase II is presented in Section 2.2.

CHAPTER 2.0

EXPERIMENTAL APPROACH

To address the objectives presented in Section 1.2, a two-phase experimental program was undertaken. Phase I, which was carried out during the first 11 months of the study (August 2004 through June 2005), focused on assessing and comparing the operational performance of an external and a submerged membrane AnMBR in terms of achievable contaminant removal and membrane filtration characteristics. A description of the experimental conditions used in Phase I is presented in Section 2.1.

Phase II, which was completed during the last seven months of the study (September 2005 through March 2006), focused on a further assessment of the membrane filtration characteristics in an AnMBR. A description of the experimental conditions used Phase II is presented in Section 2.2.

2.1 Phase I — Experimental Approach

Each AnMBR used in Phase I was constructed at bench scale (20–30 L liquid working volume). The designs for both the external and the submerged membrane systems were based on information gained from previous operation of aerobic MBRs at the University of British Columbia (UBC) pilot plant and with suggestions provided by membrane manufacturers. Figure 2-1 and 2-2 present schematics of the external and submerged membrane AnMBR systems, respectively.

2.1.1 Bench-Scale AnMBRs

2.1.1.1 External MBR

The external membrane AnMBR system (Figure 2-1) consisted of a completely mixed anaerobic bioreactor, equipped with a timer-controlled feed pump and a continuously operated progressing cavity pump, which recirculated anaerobic mixed liquor to two parallel ultrafiltration membrane trains. The feed pump was operated in a semicontinuous mode, and the flow rate through the feed pump was adjusted such that the required influent flow could be added during a 5-min period, every 15 min. The flow through the progressive cavity pumps was set to achieve a bulk cross-flow velocity (CFV) of approximately 2 m/s through the membranes. Permeate generated in the membrane modules was either discharged from the system, or returned to the bioreactor, so that the overall rate of permeate discharge could be matched to the influent flow rate. A set of liquid level sensors located outside the bioreactor controlled the permeate discharge and maintained the bioreactor liquid level within high and low alarm set-points. Temperature and pH were measured continuously and bioreactor off-gas was passed through a collapsible buffer bag to prevent large pressure swings, and then through a wet-tip gas meter for measurement of the gas production rates. The system was operated under constant TMP and variable permeate flux conditions. The TMPs were controlled manually by adjustment of a throttling valve downstream of each membrane module. The operating TMPs for the external membrane AnMBR are discussed in Section 3.1.6. The external membrane AnMBR system was initially started up without temperature control. However, the unit was later retrofitted with an internal coil for cooling water, which was used in conjunction with a thermostat to provide cooling of the bioreactor for temperature control. The hydraulic retention times (HRTs) and OLRs for the external membrane AnMBR are discussed in Section 3.1.1. The membranes were chemically cleaned when the permeate flux decreased to unsustainable levels as recommended by the membrane manufacturer (see Section 2.1.3).

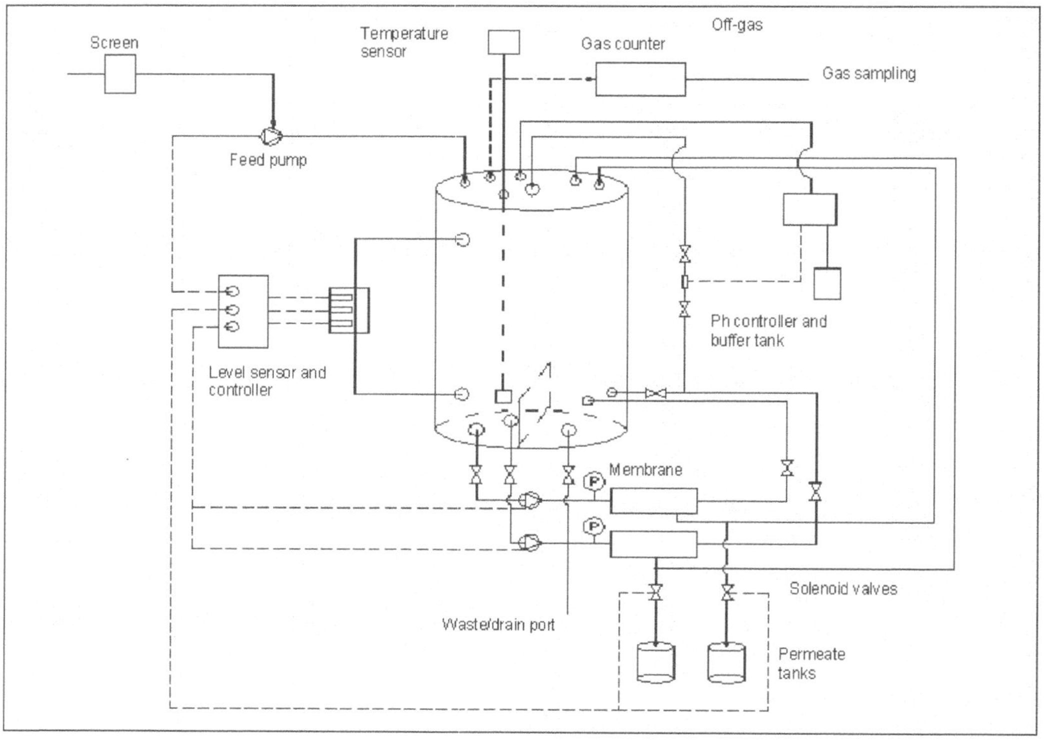

Figure 2-1. Schematic of External Membrane Anaerobic Bioreactor System.

Three different, commercially available membrane units were used at various times during Phase I of the study. The specifications for the membranes, as supplied by the manufacturers, are summarized in Table 2-1.

Fouling of the PCI membrane became so extensive that it was permanently removed from service on day 151 of Phase I (see Section 3.1.6). As presented in Table 2-1, the PCI membrane consisted of five tubular membranes in a parallel configuration. It is likely that the PCI membrane system failed because it was not possible to maintain a sufficiently high CFV through all five parallel tubular membranes. To sustain the higher permeation rates needed to achieve the low HRTs and higher OLRs that were targeted in the study, a second Membralox single tubular unit was added in series with the first unit on day 128 of Phase I. In addition, a parallel filtration system consisting of four Koch single tube membrane units operated in series was also installed on the external MBR on day 193 of Phase I. The intent of series operation of the single tube membrane units, rather than multiple membrane tubes in parallel operation, was to permit the system to achieve the required CFVs in the membrane systems while minimizing the mixed liquor recirculation pumping rates that were required to achieve these CFVs.

Table 2-1. Membranes Used in External Membrane AnMBR.

Parameter	PCI	Membralox	Koch
Description	A5-Mini module; polymeric membrane, Tube shell in ABS	Inorganic (ceramic) membrane in stainless steel shell.	PVDF co-polymer, negatively charged
Tube configuration	Five tubes of 6 mm diameter and 1,200 mm length	Single tube of 6 mm diameter and 300 mm length	Single tube of 12.5 mm diameter and 1,500 mm length*
Pore Size (μm)	0.1	0.05	120,000 daltons
Total Surface Area (m^2)	0.1	0.055	0.1
Overall Dimensions	1200 mm long x 25 mm diameter	300 mm long x 25 mm diameter	1500 mm long x 40 mm diameter

* modified from a five-tube system

2.1.1.2 Submerged MBR

The submerged membrane AnMBR system was constructed to provide a similar working volume to that of the external membrane system and the unit was equipped with similar monitoring and control equipment to that discussed above for the external membrane AnMBR. The submerged membrane AnMBR system (Figure 2-2) consisted of a completely mixed anaerobic bioreactor, equipped with a timer-controlled feed pump and two continuously operated permeate pumps connected to the submerged membranes. The feed pump was operated in a semi-continuous mode, and the flow rate through the feed pump was adjusted such that the required influent flow could be added during a 5-min period, once every 15 min. Permeate generated in the membrane modules was either discharged from the system, or returned to the bioreactor, so that the overall rate of permeate discharge could be matched to the influent flow rate. A set of liquid level sensors located outside the bioreactor controlled the permeate discharge and maintained the bioreactor liquid level within high and low alarm set-points. Temperature and pH were measured continuously and bioreactor off-gas was passed through a wet-tip gas meter for measurement of the gas production rates. The membrane modules in the submerged AnMBR were equipped with gas sparging systems to control fouling and biomass attachment to the membrane surfaces. Head-space gas was withdrawn from above the liquid level in the reactor and was pumped to the gas sparging systems located beneath each membrane unit. The sparging intensity for each membrane module was set to approximately 9 L/m^2·min. A collapsible gas buffer bag was installed in the reactor off-gas system to prevent the generation of a vacuum in the off-gas system. The submerged membrane AnMBR was operated under constant permeate flux and variable TMP conditions. A relatively constant flux was maintained using a variable speed peristaltic pump. However, as presented in Section 3.2.1.6, a constant permeate flux could not be consistently maintained due to extensive fouling. Each membrane had its own permeate pump, (Masterflex L/S peristaltic pump). The permeate pumps were operated with a 5 min on, 1 min off sequence to introduce relaxation cycles to enhance the permeate flux. The HRTs and OLRs for the submerged membrane AnMBR are discussed in Section 3.2.1.1. The membranes were chemically cleaned when the TMP increased to a value greater than approximately 6–8 psi (41–55 kPa) as recommended by the membrane manufacturer (see Section 2.1.3).

The submerged membrane system was initially designed for use with a US Filter/Memcor (Memcor) membrane module, as well as an Ionics/Mitsubishi (Mitsubishi) membrane module. However, the Mitsubishi membrane module was never used in this application due to its fragile nature. Ultimately, a segment of a Toray flat sheet membrane system was installed to operate in parallel with the Memcor hollow fibre unit. The specifications for the membranes, as supplied by the manufacturers, are summarized in Table 2-2.

Table 2-2. Membranes Used in Anaerobic Submerged MBR.

Parameter	Memcor	Toray
Description	Hollow fiber, PVDF polymeric	*Flat sheet, PVDF polymeric
Pore Size (µm)	0.08	0.08
Total Surface Area (m^2)	0.5	0.142
Dimensions	75 mm diameter x 550 mm long	150 mm x 235 mm

* Modified from a full-scale membrane module/sheet

As for the external membrane system, additional filtration surface area was eventually required to achieve the feed rates targeted in the study. Therefore, a second Memcor module was installed on day 241 of Phase I. Additional surface area was provided when a Toray flat sheet membrane was installed on day 195 of Phase I.

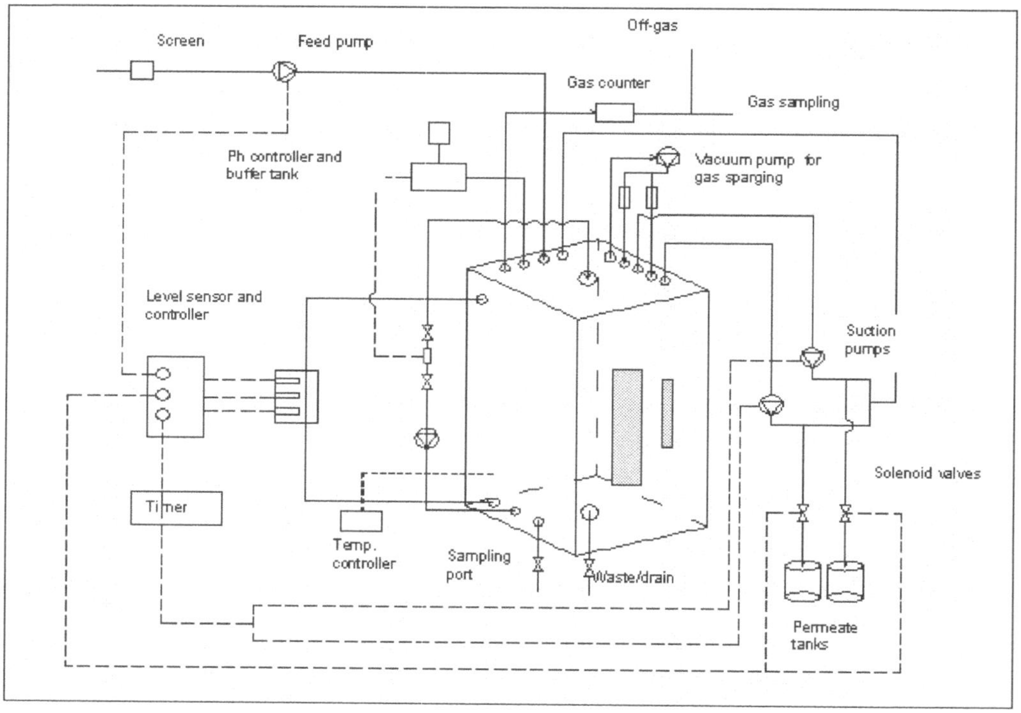

Figure 2-2. Schematic of the Submerged Membrane Anaerobic Bioreactor System.

2.1.2 Start-Up and Acclimatization

2.1.2.1 External Membrane AnMBR

The process of acclimatization of an anaerobic culture started at the beginning of August, 2004 on the external membrane reactor. Initially, the preferred approach to process seeding was to use anaerobic sludge from an existing anaerobic treatment system. The sludge chosen for this task was obtained from a full-scale upflow anaerobic sludge blanket (UASB) reactor that was treating pulp and paper mill wastewater. Sludge was also obtained from the UBC pilot-scale enhanced biological phosphorus removal system, though it was initially aerobic. The initial seed used was thus a combination of aerobic and anaerobic sludge mixed at a ratio of 20:1. The aerobic sludge was left idle initially for dissolved oxygen depletion prior to the anaerobic sludge addition.

The acclimatization process was started by operating the external membrane system in a fed batch mode during the first 12 days of operation. During that time, 6 L of sewage was added every day while the same volume of permeate from the membrane was removed to keep a constant water level and a reactor liquid volume of 20 L. One additional liter of anaerobic seed sludge from the pulp and paper mill UASB was added to the reactor every four days. After 12 days of operation in a batch fed mode, the reactor was switched to semicontinuous operation. The flow rates and OLRs to the system are discussed in Section 3.1.1. During the entire acclimatization period, the system operated at a temperature between 32° and 36°C, with mixed liquor pH values of about 6.5.

After 20 days of acclimatization, no gas production or COD removal was detected. As a result, the reactor was then reseeded with anaerobic digester sludge from a local municipal wastewater treatment plant. Acetate mixed with 1 L of wastewater was then added batch wise each day for 14 days to maintain consistent biogas production. Temperature, pH, COD, volatile fatty acid (VFA), and gas composition were then monitored on a regular basis. The temperature was maintained at 30°C while the pH averaged about 7.5.

Ultimately, the reactor was operated in a semicontinuous mode. The feed was screened primary clarifier effluent, with or without approximately 200 mg/L of supplemental sodium acetate. Periods of operation with and without acetate supplementation are discussed in Section 3.1.1.

2.1.2.2 Submerged Membrane AnMBR

Operation of the submerged membrane anaerobic bioreactor was initiated on October19, 2004. The start-up of the submerged membrane anaerobic bioreactor followed a similar approach as that of the external membrane system. The seed for the system was the same anaerobic digester sludge from a local wastewater treatment plant as was used in the external membrane system. Once the system was purged of air with nitrogen gas, 10 L of sludge was mixed with 7 L of sewage combined with enough acetate to achieve a final acetate concentration of 300 mg/L for the reactor volume. (The volumes were chosen to obtain optimal mixed liquor suspended solids [MLSS] and liquid level.)

By the end of October, the reactor was still operating under fed batch conditions nearly identical to those used initially with the external membrane reactor. Daily addition of acetate-supplemented primary effluent was preceded by decanting of the effluent (permeate) through the membranes. As with the external membrane system, eventually the submerged reactor system was operated in a semicontinuous mode. Flow rates and OLRs to the system are discussed in Section 3.2.1.1.

2.1.3 Membrane Cleaning

A simple cleaning procedure, which consisted of sequentially cleaning the membranes with a sodium hypochlorite solution and a citric acid solution, was initially utilized. However, it was not possible to recover the permeability of the membrane with this procedure. A cleaning sequence that repeatedly alternated between sodium hypochlorite cleaning and citric acid cleaning was needed to effectively recover the membrane permeability. The cleaning procedures employed for the external and submerged membrane AnMBRs are presented below.

2.1.3.1 External Membrane AnMBR

During cleaning, the membranes were taken off-line and transferred to an external membrane chemical cleaning module which consisted of a chemical cleaner reservoir and a pump which circulated the chemical cleaner from the reservoir, through the membrane, and back to the reservoir. Any chemical cleaner that permeated through the membrane was also returned to the chemical cleaner reservoir. The chemical cleaning sequence was:

1. Rinse and filter with tap water for approximately 5 min.
2. Filter with a 300–600 mg/L bleach solution, pH 10.5 at about 40°C for 20 min.
3. Filter with tap water for approximately 5 min until pH is approximately 7.
4. Filter with a 2,000 mg/L of citric acid solution, pH 2, at about 40°C for 20 min.
5. Filter with tap water for approximately 5 min until pH is approximately 7.
6. Repeat steps 1 to 4 two additional times.

2.1.3.2 Submerged Membrane AnMBR

During cleaning, the membranes were taken off-line and transferred into a submerged membrane chemical cleaning module, which consisted of a chemical cleaner reservoir and a pump which drew permeate through the membrane and back to the reservoir. The chemical cleaning sequence was:

1. Soak in tap water for 20 min.
2. Soak in a 1,000 mg/L bleach solution at pH 10.5 while sparging with air for 10 min, then filter the bleach solution though the membrane for 10 min while sparging.
3. Soak in tap water for 5 min then filter tap water for approximately 5 min while sparging (i.e., until pH is approximately 7).
4. Soak in a 2,000 mg/L citric acid solution at pH 2 while sparging air for 10 min, then filter the citric acid solution through the membrane for 10 min while sparging.
5. Repeat steps 1 to 4 two additional times.

2.1.4 Monitoring Program

2.1.4.1 Routine Monitoring

The sampling and analysis schedule shown in Table 2-3 was developed to obtain a consistent and comprehensive evaluation of the performance of the AnMBRs. In the table, VSS represents the volatile suspended solids (VSS). The routine monitoring program began in September 2004. All the onsite readings (temperature, pH, permeate flux) were usually taken in the mornings and only once per day. Samples taken to the lab for analysis were transported in a timely fashion (usually within 20 min) or preserved in accordance with *Standard Methods for the Examination of Water and Wastewater* (APHA et al., 1999). The results from the monitoring program are presented in Chapter 3.0.

Table 2-3. Monitoring Program to Track Ongoing Process Acclimation and Performance.

Parameter	Monday	Tuesday	Wednesday	Thursday	Friday	Saturday	Sunday
Temperature	x	x	x	x	x	x	x
pH	x	x	x	x	x	x	x
Gas Count	x	x	x	x	x	x	x
Permeate flux	x		x		x		
COD	x			x		x	
VFA	x			x		x	
Gas Composition	x			x		x	
TSS/VSS	x						

All analyses were performed according to *Standard Methods* (APHA et al., 1999). The expected gas production was estimated by using the following formula:

$$\text{Expected Gas Production} = \text{Flow Rate x } \frac{(\text{Total Influent COD} - \text{Effluent COD})}{1000 \text{ x Methane Yield}} \tag{2.1}$$

where the expected gas production was in L/day, the influent wastewater flow rate was in L/day, the COD in mg/L and the methane yield was assumed to be 0.35 L/g COD (Hall, 1992). As presented in Sections 3.1.3 and 3.2.2.3, methane typically accounted for approximately 95% of the total volume of biogas produced.

2.1.4.2 Periodic Monitoring

In addition to routine monitoring for the parameters outlined in Section 2.1.4.1, additional parameters were monitored on a periodic basis. A description of these parameters is presented below.

Particle Size Distribution

Periodic measurements of the mixed liquor particle size distributions were made using a laser diffraction particle size analyzer (Malvern Instruments, Hydro 2000S).

Off-Line Filtration Tests

Off-line tests were performed to assess and compare the membrane filtration characteristics of the mixed liquors generated in the external and submerged membrane AnMBRs and to compare these to similar data for an aerobic MBR (Zhang and Hall, 2006) operated with the same influent wastewater. The membranes used for this part of the study were single polyvinylidene fluoride (PVDF) hollow fibers with pore size of 0.04 μm and surface characteristics that are hydrophilic and negatively charged. These membranes were relatively similar in geometry and material to the Memcor membranes used in the submerged AnMBR discussed above, but were provided by a separate membrane manufacturer. The off-line tests were performed during the last months of Phase I of the study (April and May 2005).

The filtration set-up shown in Figure 2-3 was used for the off-line filtration studies. The bench-scale testing apparatus consisted of a 2 L reactor that supported four strands of hollow fiber membranes (total surface area: 0.004 m^2) and a gas sparging system. A digital pressure gauge was installed on the permeate line to measure the evolution in the TMP with time. The filtration system was driven by a variable speed Masterflex positive displacement pump.

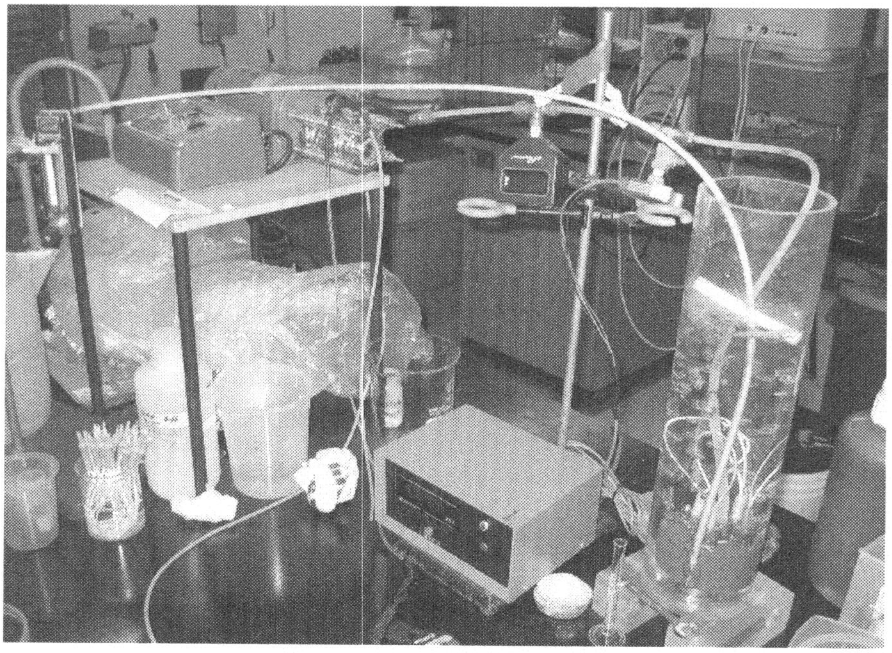

Figure 2-3. Off-Line Membrane Filtration Apparatus.

To compare the membrane filtration characteristics of the mixed liquors generated in the external and submerged membrane AnMBRs and to compare these to similar data for an aerobic MBR operated with the same influent wastewater, a series of five experiments, as presented in Table 2-4, was performed. A detailed description of the aerobic MBR system was provided by Zhang and Hall (2006). In order to compare the filtration characteristics of different mixed liquors, similar experimental conditions were adopted for all off-line membrane filtration testing. The permeation rate was set to 2.2 mL/min, which was approximately equal to a membrane permeate flux of 33 $L/m^2 \cdot hr$. The air/gas sparging intensity was set at 1.8 L/min.

Table 2-4. Sludge Type and Concentrations Tested.

Test	Sludge Type and Concentration
1	Aerobic sludge at 6 g/L concentration.
2	Anaerobic submerged membrane MBR sludge with concentration of 7 g/L.
3	Diluted anaerobic submerged membrane MBR sludge with concentration of 5 g/L.
4	Anaerobic external membrane MBR sludge with concentration of 14 g/L.
5	Diluted anaerobic external membrane MBR sludge with concentration of 5 g/L.

The dependence of permeate flux on the applied TMP can be expressed using Equation 2.2:

$$J = \frac{TMP}{\mu(Rm + Rc + Rp + Ra)} \qquad (2.2)$$

where J is the permeate flux, TMP is the applied trans-membrane pressure, and μ is the permeate viscosity. *Rm* is the intrinsic membrane resistance, *Rc* is the resistance due to the cake layer formed on the membrane surface, *Rp* is the resistance due to plugging in membrane pores, and *Ra* is the resistance due to adsorption of organic species. For the present study, all resistance values are given in units of m^{-1}.

Total resistance encountered during the filtration of sludge is due to the resistance of the membrane (*Rm*), plus the resistance invoked by the activated sludge (*Rs*); which in turn is composed of the three resistances described above: *Rc*, *Rp*, and *Ra*. *Rm* can be estimated by subjecting new membrane modules to filtration with distilled de-ionized water (DDW). *Rs* can be obtained from the differences observed between filtration tests conducted with DDW and mixed liquor. In a filtration process, *Rc* can be eliminated with the removal of bulk sludge along with relaxation and gas sparging. *Rp* is usually reduced by the process of back-flushing. *Ra* is hydraulically irreversible and can only be eliminated through chemical cleaning. The relationship among these resistances is illustrated in Figure 2-4.

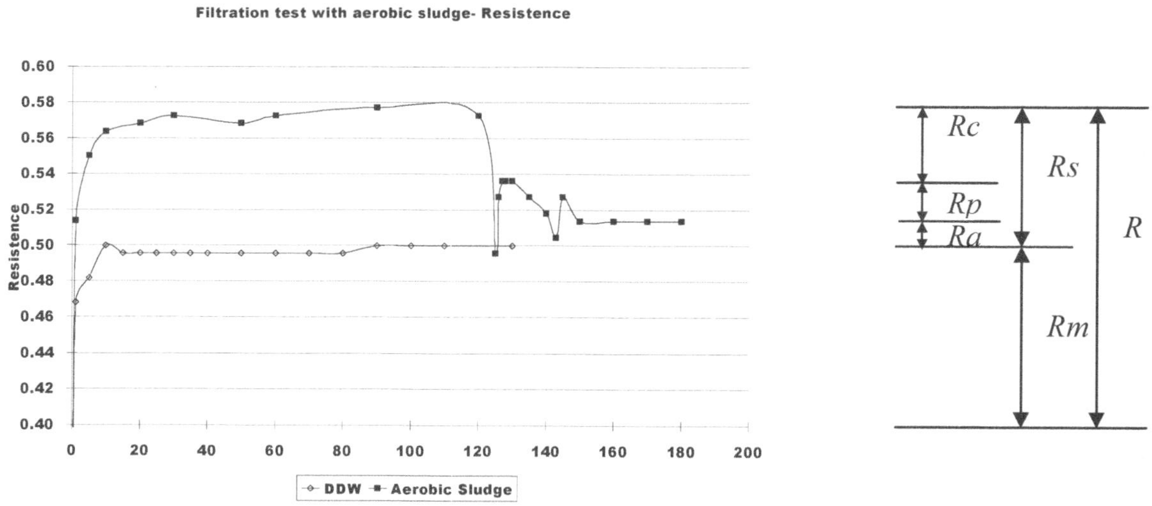

Figure 2-4. Typical Results from Off-Line Filtration Tests.

The fibers used in the off-line tests for the submerged AnMBR mixed liquor were also subjected to SEM-EDX analysis.

Scanning Electron Microscope — Energy Dispersive X-ray Analysis

The samples (membrane fiber loops from off-line tests for the submerged AnMBR mixed liquor) for the scanning electron microscope (SEM) — energy dispersive x-ray (EDX) microanalysis were prepared following the SEM protocol developed by the UBC Electron Microscope Imaging Laboratory. After fixation, dehydration, drying, and gold coating, the fouled and virgin (control) fibers were placed under a Hitachi S4700 SEM for high-resolution imaging.

2.1.5 Influent Wastewater Characteristics

The influent to the AnMBRs was municipal sewage pumped from a trunk sewer that services portions of UBC and the adjacent housing neighborhoods. Sewage is pumped approximately four times per day to two large surface-storage tanks that are gently mixed to prevent extensive suspended solids (SS) sedimentation. Stored wastewater is then pumped to a pilot-scale primary clarifier. For part of the present study, primary clarifier effluent was stored in tanks and supplemented with approximately 200 mg/L of acetate, before feeding to the anaerobic reactors. Acetate supplementation was discontinued on day 225 of Phase I (see Section 3.1.2).

The measured characteristics of the municipal wastewater during Phase I, without acetate supplementation, are summarized in Table 2-5. The total COD of the primary clarifier effluent averaged about 300 mg/L, with a soluble COD content of about 130 mg/L. The SS content of the wastewater was about 140 mg/L, with a high volatile content. Average nutrient concentrations of 28 mg NH_4-N/L and 2.8 mg PO_4-P/L were typical of municipal wastewaters.

Table 2-5. Phase I Influent Municipal Wastewater Characteristics.

Parameter	Average	Standard Deviation
COD (mg/L)	298	60
Soluble COD (mg/L)	131	40
COD plus acetate (mg/L)	520	101
Soluble COD plus acetate (mg/L)	334	54
TSS (mg/L)	137	
VSS (mg/L)	137	
NH4-N (mg/L)	28.5	4.2
PO4-P (mg/L)	2.8	0.5

2.2 Phase II — Experimental Approach

As presented in Chapter 3.0, slightly less fouling was observed using external inorganic membranes in an AnMBR. However, based on discussion with the PSC and a review of relevant literature, it was concluded that the high capital and operating costs associated with external inorganic membranes outweighed the slightly lower fouling tendency of these systems. Therefore, only the submerged membrane AnMBR was considered for Phase II of the study.

2.2.1 Bench-Scale AnMBR Start-Up and Acclimatization, Routine Monitoring, and Influent Wastewater Characteristics

The submerged membrane AnMBR used in Phase II was similar to that described in Section 2.1.1.2 for Phase I. However, only Memcor membrane modules were used. The same start-up and acclimatization procedure adopted for the submerged membrane AnMBR in Phase I was used in Phase II. The routine monitoring program outlined in Section 2.1.4.1 was modified slightly as presented in Table 2-6. The characteristics of the municipal wastewater, without acetate supplementation, measured during Phase II, are summarized in Table 2-7. In general, the characteristics of the municipal wastewater used during Phase II were similar to those of Phase I (see Section 2.1.5). For all of Phase II, the influent wastewater to the submerged AnMBR was supplemented with acetic acid to achieve an influent acetic acid concentration of approximately 200 mg/L. The supplementation with acetate was used to ensure that the robust anaerobic conditions could be maintained in the bioreactor without reducing the concentrations of any wastewater components that might affect membrane permeability.

Table 2-6. Monitoring Program to Track Ongoing Process Acclimation and Performance.

Parameter	Monday	Tuesday	Wednesday	Thursday	Friday	Saturday	Sunday
Temp	X	X	X	X	X	X	X
pH	X	X	X	X	X	X	X
Gas Count	X	X	X	X	X	X	X
Flux	X	X	X	X	X	X	X
COD	X		X		X		
VFA	X		X		X		
Gas Composition	X		X		X		
TSS/VSS	X						

Table 2-7. Phase II Influent Municipal Wastewater Characteristics.

Parameter	Average	Standard Deviation
COD (mg/L)	252	58
Soluble COD (mg/L)	133	36
COD plus acetate (mg/L)	466	58

2.2.2 Periodic Monitoring

In addition to routine monitoring for the parameters outlined above, additional parameters were monitored on a periodic basis. A description of these parameters is presented below.

2.2.2.1 On-Line Filtration Tests

The experimental conditions (i.e., permeate flux and hydrodynamic conditions) used during the off-line filtration tests (Section 2.1.4.2) performed during Phase I of the present study, were significantly different from the conditions applied to the bench-scale AnMBRs. Both the operating permeate flux and the system hydrodynamics (i.e., sparging intensity and system configuration), which greatly affect mass transfer and, therefore, fouling at the membrane surface, were different in the off-line tests and the bench-scale AnMBRs. In addition, the off-line tests only lasted a few hours, while the membranes in bench-scale AnMBRs were operated for a number of months. To assess the membrane filtration characteristics of mixed liquors under more relevant permeate flux and hydrodynamic conditions, on-line filtration tests were conducted during Phase II of the present study. On-line filtration tests were performed on both anaerobic and aerobic mixed liquors.

For the on-line tests, two cleaned Memcor membrane modules were installed in the submerged membrane AnMBR described in Section 2.1.1.2, and a nearby pilot-scale anaerobic-anoxic-aerobic MBR (hereafter referred to as aerobic MBR), on days 133 and 135, respectively, of Phase II of the study. The influents to the AnMBR and the aerobic MBR were both drawn from the same wastewater source. A detailed description of the aerobic MBR pilot plant was provided by Zhang and Hall (2006). The membrane modules in the AnMBR and aerobic MBR systems were operated with the same permeate flux (15 $L/m^2 \cdot hr$), relaxation cycles (5 min on, 1 min off) and sparging intensities (9 L/min per module). The operating permeate flux selected for the on-line filtration tests was substantially higher than that used prior to the tests (3 to 5 $L/m^2 \cdot hr$; see Section 3.2.2.5). A higher flux was selected for the on-line tests so that both the AnMBR and aerobic MBR systems would be subjected to commercially relevant operating conditions. Discussions with the membrane supplier (US Filter) indicated that, an operating permeate flux in the order of 15 to 30 $L/m^2 \cdot hr$ is typical for full-scale aerobic MBR systems treating municipal wastewaters. A lower operating permeate flux had been adopted during Phase II prior to the on-line tests since the results from Phase I of the present study indicated that operating at a permeate flux greater than approximately 5 $L/m^2 \cdot hr$ results in rapid fouling under anaerobic treatment conditions (see Section 3.2.1.6).

To assess the fouling propensities of each mixed liquor, the TMPs were monitored over time. In addition, fibers were taken from the Memcor membrane modules at the start (days 133 and 135 for the AnMBR and the aerobic MBRs, respectively) and toward the end of the on-line fouling study (day 156) for SEM-EDX microanalysis. Fibers were also taken from the partially fouled Memcor membrane modules towards the end of the on-line fouling study (day 156) for chemical analysis of extracted foulants. The fibers were cut directly from the membrane modules and the ends of the cut fibers that remained on the modules were sealed with epoxy.

2.2.2.2 Scanning Electron Microscope-Energy Dispersive X-Ray Analysis

The SEM-EDX microanalysis used in Phase II was similar to that in Phase I, with the exception that the analysis was performed on fibers taken from the on-line filtration tests described above.

2.2.2.3 Extraction of Membrane Foulants

The fouled membranes were subjected to basic extraction and acidic extraction. For basic extraction, the membrane fibers from the on-line filtration tests described above were soaked in 9.0 mLof 0.1M NaOH solution and sonicated (Aquasonic Model 550HT, VWR) at 40°C for 1 hour. The fibers were then removed and the solution (i.e., extract) was then filtered through a 0.45 μm filter paper and then neutralized with 3 mL of 0.15-M H_2SO_4. For all tests, the extractions were done in triplicate (using three fibers). For acidic extraction, the procedure was similar to that described for basic extraction except that the membrane fibers were soaked in 9 mL of 0.5-M H_2SO_4 and subsequently neutralized with 3 mL of 3-M NaOH solution.

2.2.2.4 Chemical Analysis of Extracted Fouling Material

The extracted samples were analyzed for total Kjeldahl nitrogen (TKN), total phosphorus (TP), and TOC as described in *Standard Methods* (APHA et al., 1999). The extracted samples were analyzed for soluble microbial product (SMP) content, in terms of carbohydrate, protein, and humic substance content. The analysis of the carbohydrate content of the extracts was completed following the method suggested by Frølund et al. (1996) and the analysis of protein and humic substance content of the extracts was determined following the modified Lowry method (Jahn and Nielsen, 1995).

SMP analysis was also performed on the mixed liquor for both the AnMBR and the aerobic MBR on days 145 and 160 of Phase II.

2.2.2.5 Refractory Organic Content

As discussed in Sections 3.1.2 and 3.2.1.2, a substantial amount of the influent COD was estimated to be refractory to anaerobic treatment. The amount of refractory COD was estimated by assuming that the membrane permeate COD was composed of only two categories of COD: volatile acid COD (or biodegradable COD), and nonbiodegradable COD (or "refractory" COD). The difference between a few measurements of $CBOD_5$ and the COD was also used as an indicator of the amount of refractory material present in the AnMBR influent and effluent during Phase II of the study (see Section 3.2.2.2). Both COD and $CBOD_5$ were analyzed as described in *Standard Methods* (APHA et al., 1999).

CHAPTER 3.0

Assessment of the Operational Performance of an External and a Submerged Membrane AnMBR

The overall performance of AnMBRs with different membrane filtration systems was assessed in terms of achievable contaminant removal and filtration rates. The overall performance assessed during both Phase I and Phase II of the study is presented in this chapter.

3.1 External Membrane AnMBR (Phase I)

3.1.1 Description of Operating Conditions

The external membrane anaerobic bioreactor was put into operation on August 31, 2004. Following an initial period of fed batch operation and re-inoculation for establishment of a viable and active anaerobic biomass, semicontinuous flow operation was initiated on operating day 11 at a nominal HRT of 67 hours. Over the subsequent 300 or so days of operation, feed rates were increased and the bioreactor liquid volume was reduced somewhat, such that ultimately, a nominal operating HRT of approximately two hours was reached (Figure 3-1). Near day 300, the extent of fouling of the membranes was such that the desired feed rates could not be maintained. Consequently, the operation of the MBR was terminated for extensive membrane cleaning and for servicing of the pumps and temperature control system.

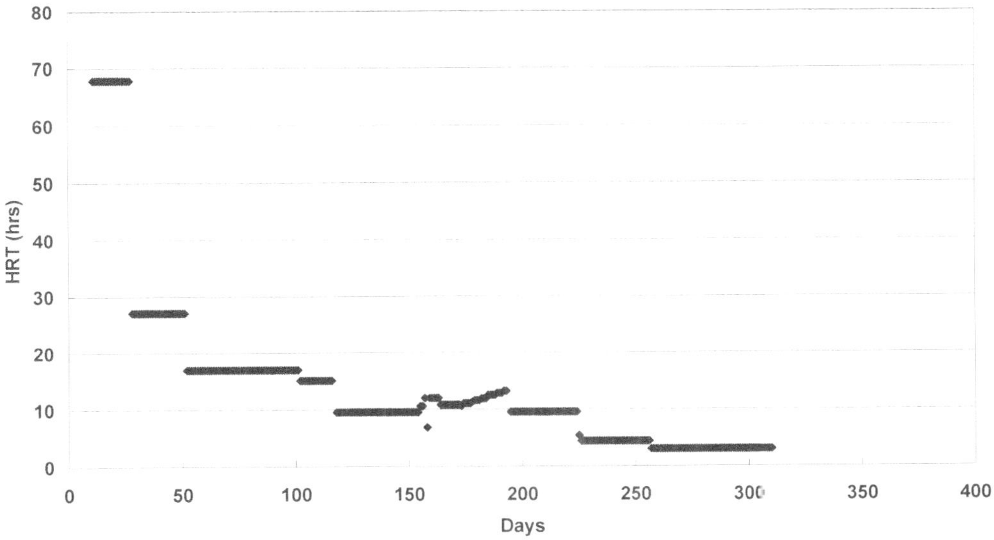

Figure 3-1. Nominal (Empty Bed) HRTs in the External Membrane AnMBR.

The process OLRs (based on bioreactor empty bed volume) associated with the feed rates described above, are shown in Figure 3-2. Prior to day 225, the AnMBR feed was supplemented by the addition of approximately 200 mg/L of acetate and OLRs of 1.0 to 1.5 kg COD/m^3·day were being achieved. Following the termination of acetate supplementation, influent flow rates of un-supplemented wastewater were increased and OLRs of 1.5 to 2.5 kg COD/m^3·day were typical. The achievable OLR was limited by the permeate flow that could be sustained through the membrane component of the AnMBR. For the external membrane AnMBR used in the present study, the maximum OLR that could be achieved was approximately 2.5 COD/m^3·day.

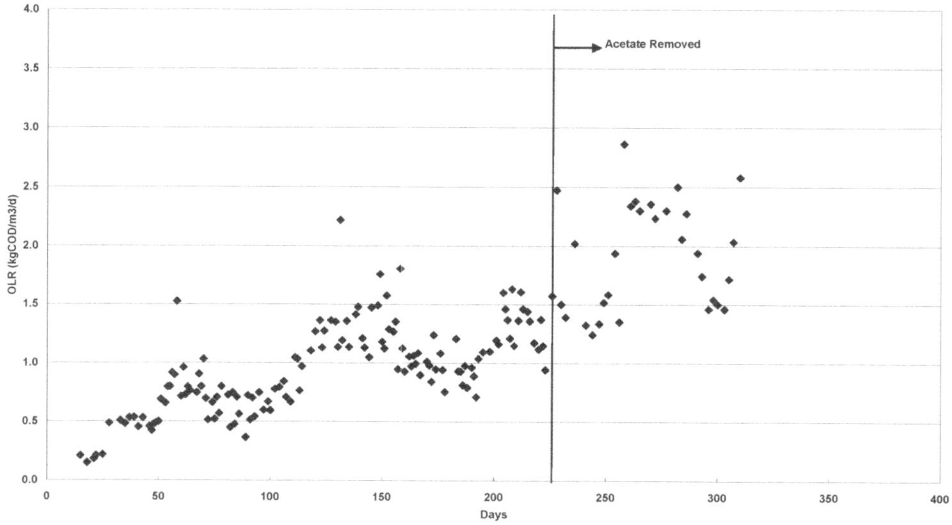

Figure 3-2. OLRs in the External Membrane AnMBR.

Initially, the operating temperature in the external AnMBR was not controlled. The energy added to the system by the extensive mixed liquor recirculation to the membrane units, resulted in relatively high operating temperatures of 30–35°C, that were considered to be satisfactory for an anaerobic bioreactor that was undergoing start-up (Figure 3-3). However, when operating temperatures exceeded 40°C, it was recognized that cooling would be needed and a thermostatically controlled cooling coil was installed in the bioreactor. The cooling system was used to control operating temperatures at approximately 30°C between days 150 and 200 and then at 25°C thereafter. A failure of the temperature control system near the end of the study resulted in loss of temperature control and operating temperatures became variable over the range from 25° to 40°C.

Figure 3-4 shows the measured pH values in the external AnMBR influent and in the bioreactor. Small quantities of sodium bicarbonate were added to the feed as required to maintain the minimum bioreactor pH (between 6.6 and 6.8). In the last part of the study (after day 200) volatile acids concentrations were observed to increase across the bioreactor, resulting in bioreactor pH readings that were slightly lower than those measured in the influent.

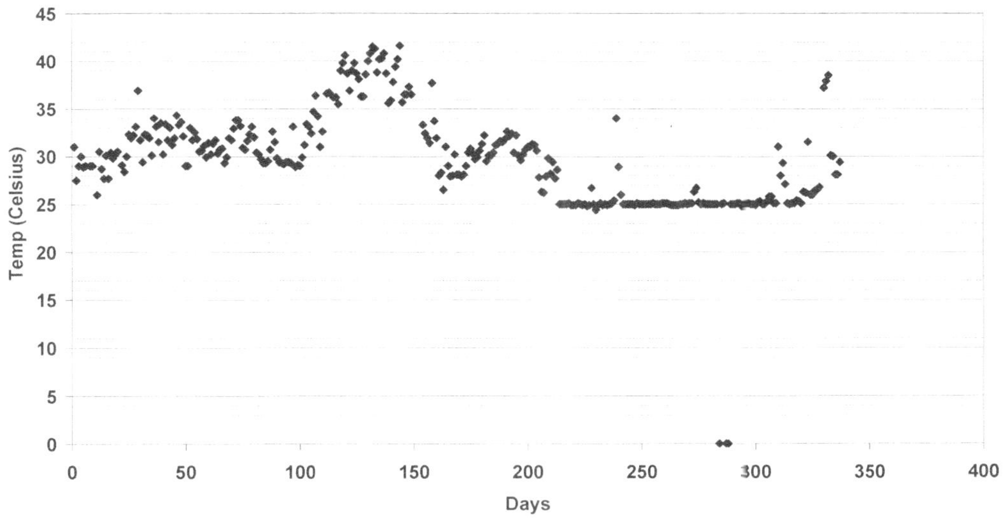

Figure 3-3. External AnMBR Temperature.

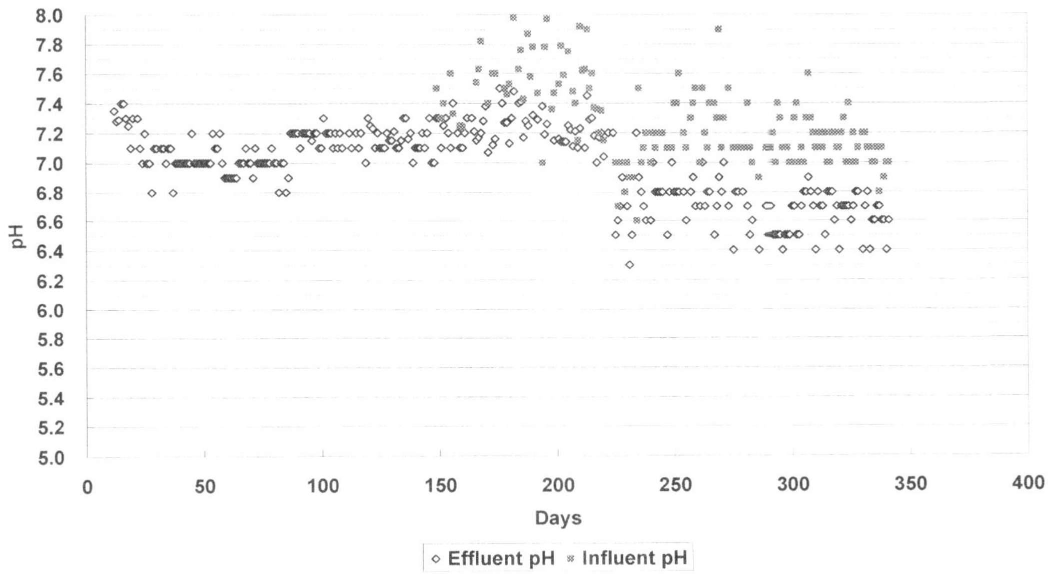

Figure 3-4. Influent and Effluent (Permeate) pH for the External Membrane AnMBR.

3.1.2 Treatment Performance

Figure 3-5 summarizes the results of influent and permeate (effluent) COD measurements completed during the periods of operation with, and without, acetate supplementation. With acetate supplementation (i.e., start of Phase I to day 224), the external membrane AnMBR influent contained an average of approximately 520 mg COD/L, of which approximately 330 mg COD/L was soluble, while without acetate supplementation (i.e., day 225 to end of Phase I), the external membrane AnMBR influent contained an average of approximately 290 mg COD/L, of which approximately 130 mg COD/L was soluble.

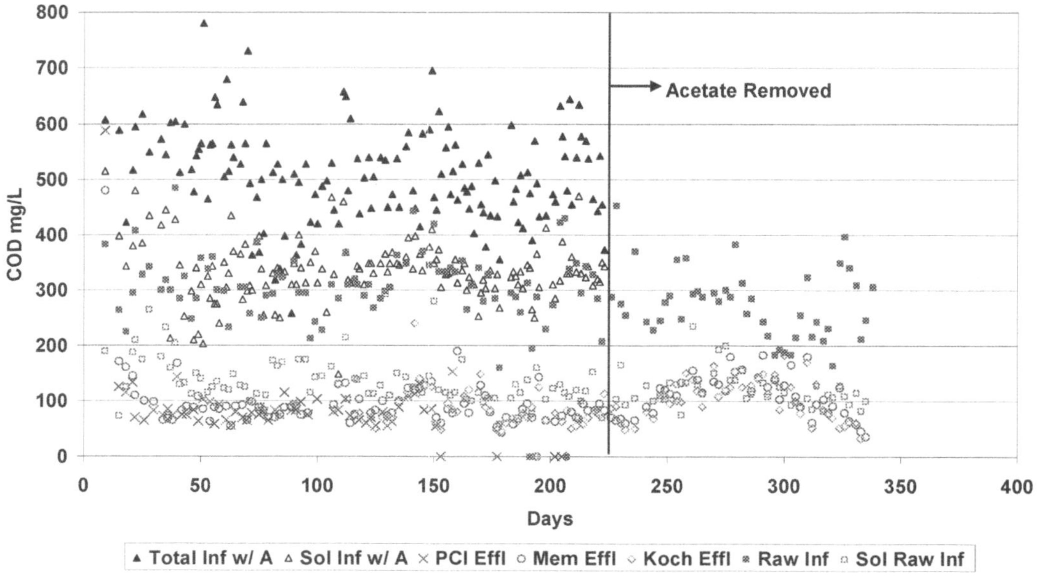

Figure 3-5. Total and Soluble COD Concentrations in Influent and Permeates for the External AnMBR System.

Table 3-1 indicates that, during operation with acetate supplementation, the average measured removal efficiencies for total and soluble COD were 82% and 73% respectively. Without acetate supplementation, total COD removal efficiencies were reduced to 57% on average. Soluble COD removal efficiencies decreased even more, to an average of 6%. Furthermore, as indicated in Table 3-1, the standard deviation on this average was quite high, such that many individual measurements of soluble COD removal efficiencies returned values that were close to or below zero.

Table 3-1. Measured COD Removal Efficiencies During the Periods of Operation with and without Acetate Supplementation.

Parameter	Average	Standard Deviation
With Acetate Supplementation		
Total COD Removal (%)	82	8
Soluble COD Removal (%)	73	11
Without Acetate Supplementation		
Total COD Removal (%)	57	14
Soluble COD Removal (%)	6	30

The nature of the effluent (permeate) COD during the experimentation was of interest. Figure 3-5 and Table 3-2 demonstrate that during the best period of operation (days 1 to 224), permeate COD averaged about 87 mg/L. It could be argued that this membrane permeate COD was composed of only two categories of COD—volatile acid COD (or biodegradable COD) and anaerobically nonbiodegradable COD (or "refractory" COD). For operation between days 1 and 224, the subtraction of VFA-COD from permeate COD then provides an indication of the "refractory" COD in the wastewater during this period. As shown in Table 3-2, the refractory COD was estimated to be about 82 mg/L on average.

Table 3-2. Average Measured Effluent (Permeate) COD, VFA-COD and "Refractory" COD Concentrations for the External Membrane AnMBR during the Periods of Operation with and without Acetate Supplementation.

Parameter	Average
With Acetate Supplementation	
Permeate COD (mg/L)	87
Permeate VFA-COD (mg/L)	5
Estimated "Refractory" COD (mg/L)	82
Without Acetate Supplementation	
Permeate COD (mg/L)	104
Permeate VFA-COD (mg/L)	32
Estimated "Refractory" COD (mg/L)	72

For the operating period between days 225 and the end of the study (without acetate supplementation), COD removal efficiencies were very low and permeate volatile acids concentrations exceeded the concentrations in the influent (Figure 3-6). This suggests that anaerobic biological activity was responsible for VFA formation via bio-acidification of wastewater COD. However, methanation of the VFAs was limited. If VFA-COD is subtracted from permeate COD for this period of operation, the estimated average concentration of refractive COD was 72 mg/L (Table 3-2), a value that is consistent with that from the acetate-supplemented period.

As a result of the filtration afforded by the membranes, suspended solid concentrations in the permeate samples were always below detection limits.

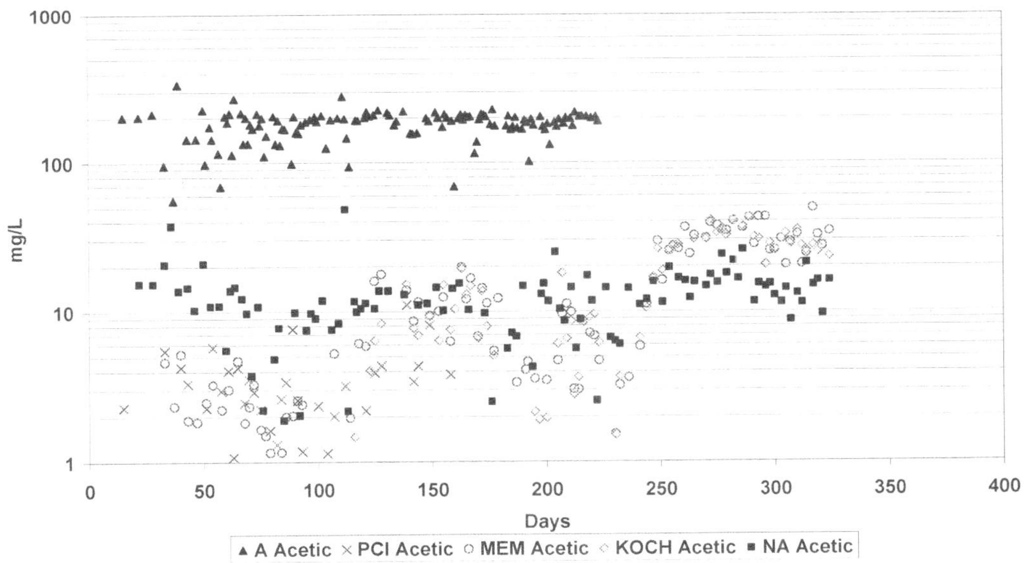

Figure 3-6. Acetic Acid Concentrations in External Membrane AnMBR Influent and Permeates.

3.1.3 Biogas Production

With an anaerobic biological wastewater treatment system, it is usually assumed that the majority of the COD removed by treatment is converted to a biogas that contains CH_4 and CO_2. The composition of the biogas is related to a number of factors, including the process operating conditions, the characteristics of the wastewater and the influent COD concentration. Biogas methane content of 65–75% are commonly observed (Hall, 1992).

For the present study, Figure 3-7 shows the measured daily total biogas production rates for the external membrane AnMBR. Also indicated in Figure 3-7 is the expected total biogas production, calculated from the instantaneous total COD removal observed across the system and applying an assumed biogas yield of 0.35 m^3/kg COD removed (see Section 2.1.4.1).

The predicted biogas production rates were highly variable, but in general, they ranged from 6 to 12 L/d. The variability in the expected biogas production rates is due to the variability that was observed in measured influent COD concentrations. Figure 3-7 also demonstrates that the expected biogas production rates remained high even beyond day 225, when the external membrane AnMBR was operated without acetate supplementation. This is due to the fact that although total COD removal efficiency decreased from 82% to 57% after day 225, the influent flow rates were increased concurrently. Taken together, this set of conditions suggested that biogas production should still remain high, even without acetate supplementation.

Figure 3-7 also indicates that the measured biogas production rates were always lower, and in some cases substantially lower, than the expected values. On average, with acetate supplementation, the expected biogas production rate was approximately 6.6 L/d, and the measured biogas production rate was approximately 4.3 L/d, while without acetate supplementation, the expected biogas production rate was approximately 8.2 L/d, and the measured biogas production rate was approximately 1.5 L/d.

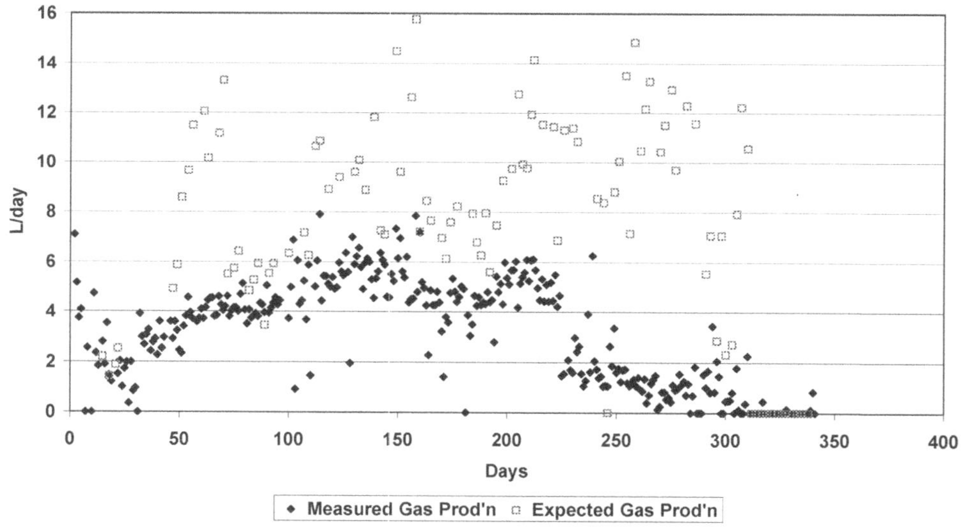

Figure 3-7. Measured and Expected Total Biogas Production Rates in External Membrane AnMBR.

There are likely at least two reasons for the discrepancy between expected and measured biogas production rates. First, when dilute wastewaters are treated anaerobically, a significant portion of the CO_2, and to a lesser extent, the CH_4, will exit the process dissolved in the effluent. The effect of this would be reduced biogas production rates and an enrichment of the CH_4 content of the biogas, relative to that of CO_2. Figure 3-8 confirms the fact that when the CH_4 and CO_2 biogas quality data are examined, the external AnMBR off-gas was unusually rich in CH_4 and relatively low in CO_2. This is a phenomenon that would not be significant with higher strength wastewaters, because the amount of biogas produced per unit volume of wastewater treated, would greatly exceed the solubility limits of CH_4 and CO_2. In addition, Figure 3-8 shows that biogas quality (CH_4 content) deteriorated near the end of the study, when process treatment efficiencies declined. Considering that the solubility of methane is approximately 17 mg/L (at 30°C and one atmosphere), the amount of CH_4 produced that is expected to exit the process dissolved in the effluent when the system is supplemented with and without acetate is approximately 16% and 41%, respectively.

The second important factor in the biogas discrepancy relates to the presence of particulate COD in the influent wastewater. All of the particulate COD is retained in the AnMBR by the membrane filtration units. The extent to which this particulate material is degraded and then fermented to CH_4 and CO_2 is not known. It would be reasonable to assume that at least some of the particulate COD is biodegradable. However, the calculation of expected biogas production assumes that all of the particulate COD is biodegradable and fermentable, leading to an overprediction of the biogas production. During the period of operation with acetate supplementation, the contribution of particulate COD to the total wastewater COD was significantly lower than during the subsequent period. Therefore, the discrepancy between expected and measured biogas rates would be expected to be smaller during this period, than after day 225, as confirmed by Figure 3-7.

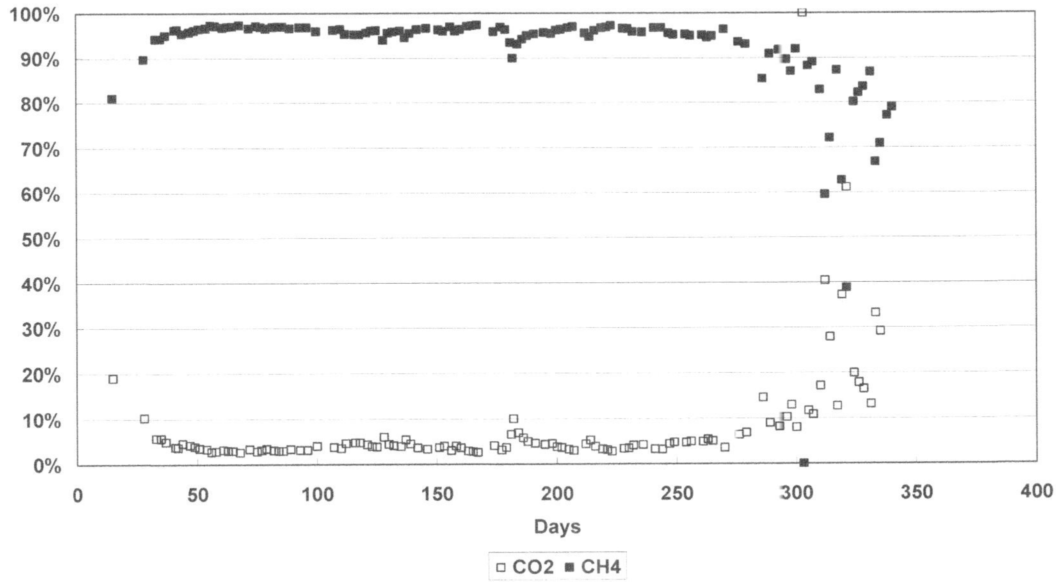

Figure 3-8. Composition of CO_2 plus CH_4 Portion of External Membrane System Off-Gas (N_2 Excluded from the Plot).

In view of the complications associated with the validation of the measured biogas production rates, it is difficult to judge the quality of the measured data. In light of the above discussion, the measured biogas data appear to be reasonably consistent with the measured COD removal data.

3.1.4 Bioreactor SS

The measured bioreactor SS concentrations in the external membrane AnMBR over the course of the study is represented in Figure 3-9. Following the seeding and biological start-up of the system, the bioreactor contained about 4,000 mg/L of total suspended solids (TSS), of which about 80% was volatile. After semicontinuous feeding of wastewater was initiated, each liter of influent introduced about 100 mg TSS into the bioreactor, all of which was retained by the membrane. Over the course of the study, the continued accumulation of these influent SS appeared to be the primary cause of the increase in MLSS shown in Figure 3-9. Following the elimination of acetate supplementation on day 225, the process feed rates were increased a number of times. As Figure 3-9 indicates, this resulted in a significant additional loading of TSS. During the last phase of the study, MLSS concentrations decreased. It is assumed that this occurred partially through endogenous decay, but more significantly, via SS loss that was known to be associated with the many membrane-cleaning events that occurred during this period.

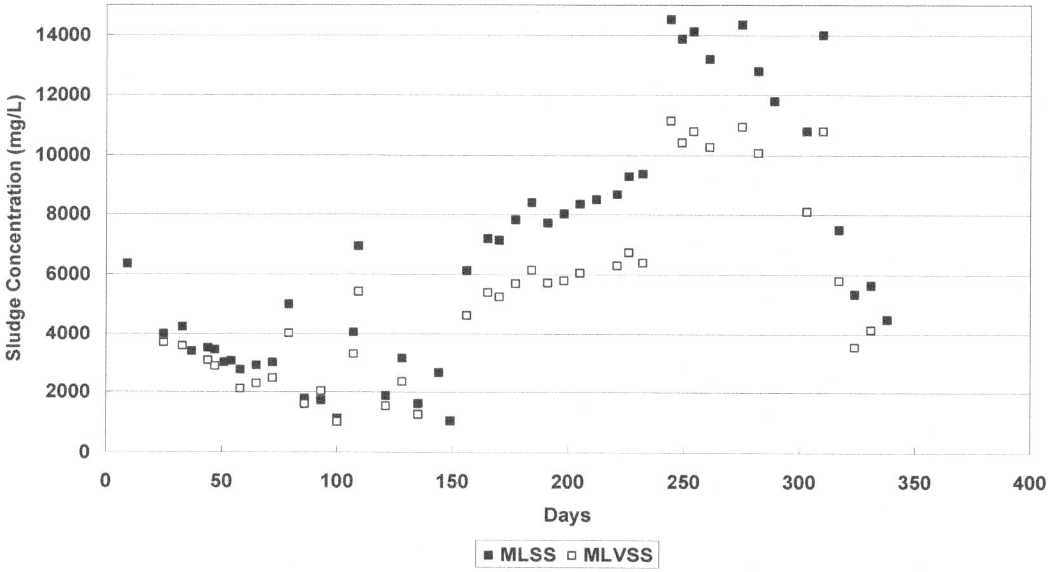

Figure 3-9. Total and Volatile SS Concentrations in External Membrane AnMBR Mixed Liquor.

3.1.5 Bioreactor SS Particle Size Distribution

Figure 3-10 shows the particle size distributions measured during the period of operation between February and May 2005 (days 178 to 262). The SS particle size distribution ranged from approximately 0.04 μm to over 0.1 mm. The lower range is consistent with the particle sizes one would expect for the membranes used (see Table 2-1). The percentage volume of particles ranging in size from 0.1 μm to over 0.1 mm was relatively constant, indicating that a much larger number of particles in the smaller size range were present in the mixed liquor.

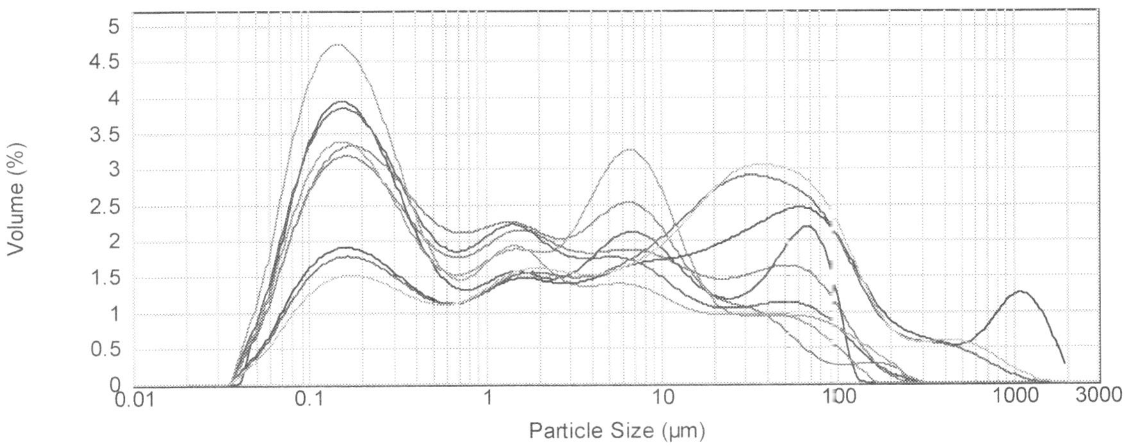

Figure 3-10. Particle Size Distributions for Mixed Liquor from External Membrane System. (Results Presented for 9 Separate Periodic Sampling Events During Phase I)

3.1.6 Membrane Flux

When the external membrane AnMBR was originally put into service, the two membrane filtration systems selected for use were the PCI polymeric unit and the US Filter Membralox ceramic filter unit. As can be seen from Figure 3-11, the flux declined steadily over the first 150 days of operation for the PCI membranes, from values near 40 L/m^2·hr, to essentially zero. Near the end of this period, the PCI membrane was cleaned twice, but satisfactory flux could not be restored. The PCI unit was then taken out of service. It is likely that the PCI membrane system failed because it was not possible to maintain a sufficiently high CFV through all five parallel tubes of membrane unit. Due to parallel configuration, the highest CFV that could be achieved through the PCI membrane system was approximately 0.5 m/s.

During the first 115 days of operation, most of the system filtration capacity was contributed by the Membralox unit, which operated at fluxes of greater than 50 L/m^2·hr. To accommodate higher feed rates planned for the AnMBR, it was decided that the Membralox TMP would be increased from 30 to 40 psi (206 to 275 kPa) to generate even greater unit flux. As noted in Figure 3-11, the higher TMP resulted in a substantial increase in flux, which then rapidly declined. A second Membralox unit was added in series to the first unit on day 128, to increase the available filtration area. With the additional surface area, membrane areal loadings were reduced by 50% and at this level, the Membralox flux was maintained with only one cleaning at day 136.

To continue addressing the objective of membrane comparison, the PCI membrane was replaced first, by one Koch unit, and then by four Koch units in series (labelled as Koch2 in Figure 3-11. With one cleaning at day 236, the Koch2 flux remained at a satisfactory level until the last two months of the program. During this latter period, the external membrane AnMBR system was subjected to a number of hardware failures that compromised the bioreactor performance. At the same time, the measured fluxes of both membranes deteriorated significantly. The Koch2 unit was cleaned six times using different approaches between days 240 and 332, but the deterioration in flux could not be arrested. By the end of the study, the combined permeation rate of the two parallel membrane systems had decreased to a level that was insufficient to maintain the target feed rates. As a result, after day 332, feed rates were relaxed to about 45 liters per day, equivalent to a nominal HRT of about 9.5 hours (data not shown).

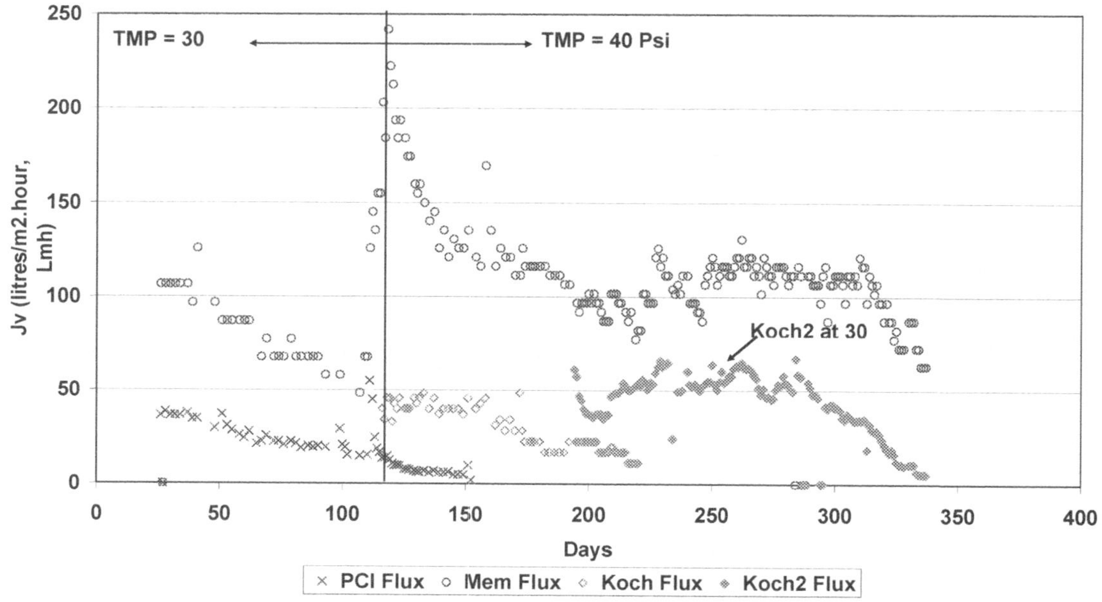

Figure 3-11. Measured Flux of PCI, Membralox and Koch Membranes in the External Membrane AnMBR System.

3.2 Submerged Membrane AnMBR

As presented in Chapter 2.0, submerged membrane AnMBRs were used during both Phase I and Phase II of the present study. The assessment of the operational performance for the submerged membrane AnMBRs used during Phase I and Phase II are presented in Section 3.2.1 and Section 3.2.2, respectively.

3.2.1 Phase I

3.2.1.1 Description of Operating Conditions

The submerged membrane AnMBR was put into operation in the fed batch mode on day 52. Batch operation continued until day 112, as the biomass became acclimatized and the hardware was debugged. On day 112, semicontinuous feeding was started at a nominal HRT of 17 hours. The feed rates were subsequently increased stepwise to reach HRTs as low as 5-10 hours. After day 300, insufficient membrane flux was available to maintain such low HRTs and as shown in Figure 3-12, the retention time increased over time to about 12 hours near the end of the study.

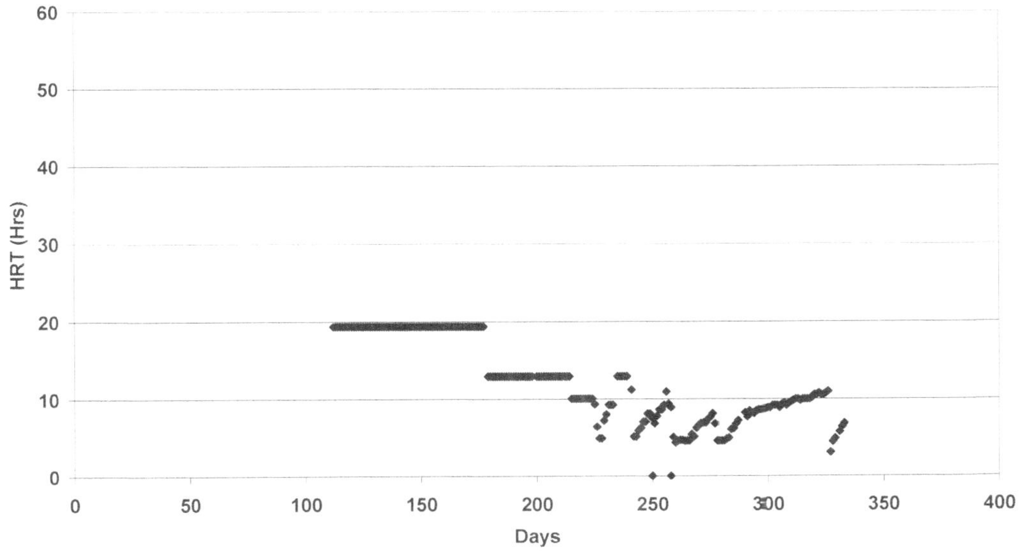

Figure 3-12. Nominal (Empty Bed) HRTs in the Submerged Membrane AnMBR.

The OLR for the submerged membrane system is shown in Figure 3-13. With influent acetate supplementation, OLRs were gradually increased from approximately 0.7 kg COD/m^3·d (empty bed volume basis) to about 1.3 kg COD/m^3·d. During the subsequent period of operation without acetate supplementation, variation in influent strength and feed rates resulted in considerable variability in the process organic loadings, but generally these were maintained in the range between 0.8 and 1.5 kg COD/m^3·d. Near the end of the study, when the treatment performance of the submerged membrane unit deteriorated, and as a result, the imposed organic loadings were reduced to about 0.5 to 0.8 kg COD/m^3·d.

Since the submerged membrane system did not rely on recycle pumps for mixed liquor recirculation to the membrane units, there was no significant heat load on the system. As a result, process temperatures were lower and more easily controlled in the submerged membrane system, relative to the external membrane system. Figure 3-14 indicates that submerged MBR temperatures were about 30°C in the first half of the study and about 25°C for the second half. As noted in Figure 3-15, the pH history of the submerged membrane systems was almost identical to that of the external membrane system.

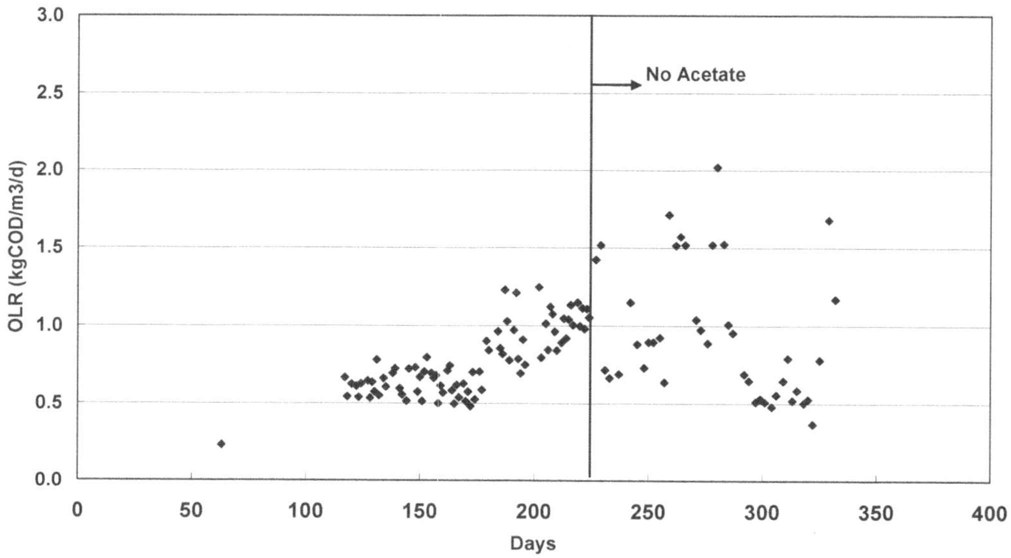

Figure 3-13. OLRs in the Submerged Membrane AnMBR.

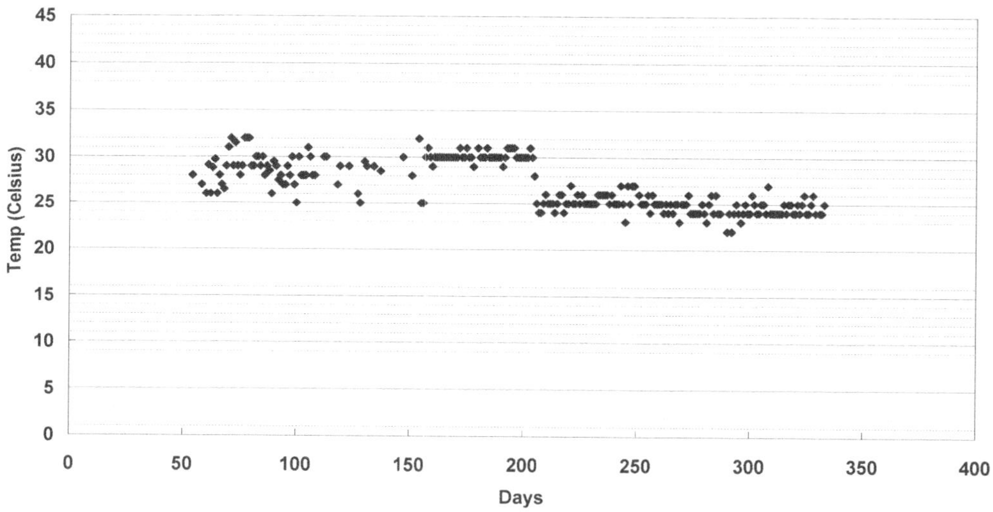

Figure 3-14. Submerged Membrane AnMBR Temperature.

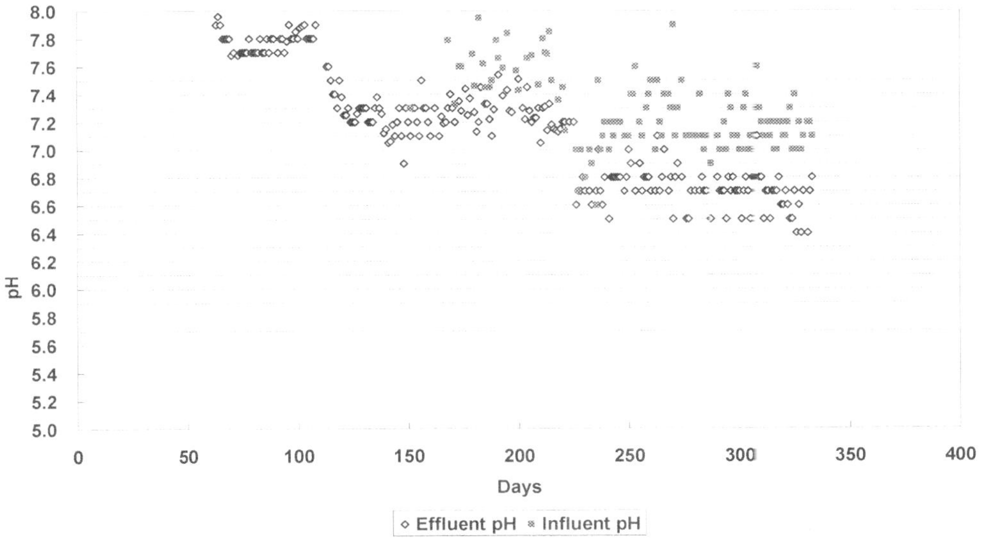

Figure 3-15. Influent and Effluent (Permeate) pH.

3.2.1.2 Treatment Performance

As noted previously, both AnMBRs were operated for periods with an acetate-supplemented influent and with an un-supplemented influent. With acetate supplementation from the start of Phase I to day 224, the submerged membrane AnMBR influent contained an average of approximately 520 mg COD/L, of which approximately 330 mg COD/L was soluble. Without acetate supplementation from day 225 to end of Phase I, the submerged membrane AnMBR influent contained an average of approximately 290 mg COD/L, of which approximately 130 mg COD/L was soluble. Table 3-3 indicates that, during operation with acetate supplementation, the average measured removal efficiencies for total and soluble COD were 76% and 67%, respectively. These performance efficiencies were similar to those observed over the same period with the external membrane system.

Without acetate supplementation, total COD removal efficiencies for the submerged AnMBR decreased to 44% on average and soluble COD removal, on average, became negative. This negative value indicates that during the last phase of the study, the operating problems associated with membrane cleaning had largely converted the bioreactor environment from a methanogenic system, to a system that was dominated by hydrolysis and acidification. During the operating period without acetate supplementation, permeate volatile acids concentrations were indeed higher than influent volatile acids concentration, suggesting that anaerobic biological activity was responsible for VFA formation via bio-acidification of wastewater COD (Figure 3-17).

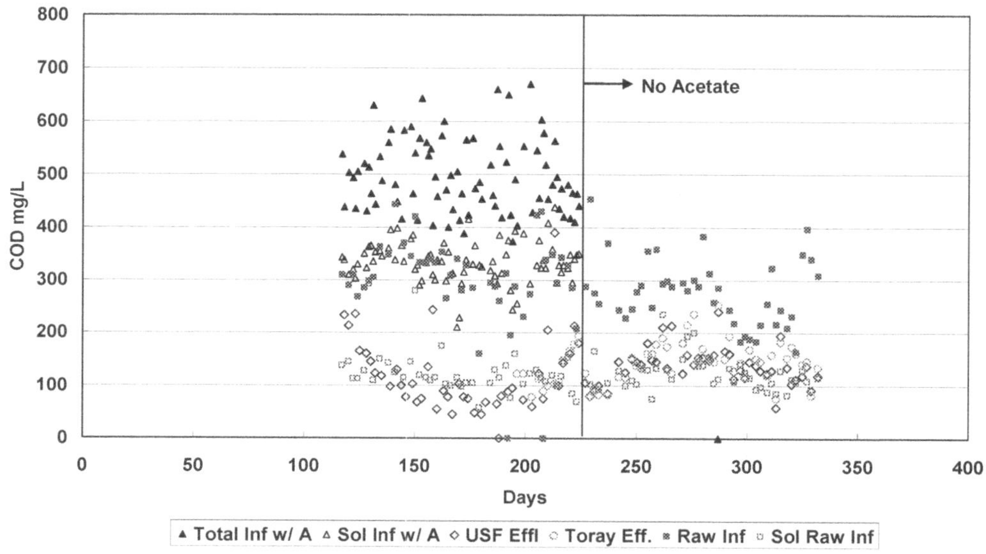

Figure 3-16. Total and Soluble COD Concentrations in Influent and Permeates for the External AnMBR System.

Table 3-3. Measured COD Removal Efficiencies in External and Submerged AnMBRs During the Periods of Operation with and without Acetate Supplementation.

Parameter	External System	Submerged System
With Acetate Supplementation		
Total COD Removal (%)	82 ± 8	76±12
Soluble COD Removal (%)	73 ± 11	67±51
Without Acetate Supplementation		
Total COD Removal (%)	57 ± 14	44±16
Soluble COD Removal (%)	6 ± 30	-21±34

Note: ±: Standard deviation of analyses

By adopting the same approach that is described above in Section 3.1.2, COD data from the submerged membrane system were used to estimate the "refractory" COD concentrations in the membrane permeates. Table 3-4 summarizes these values and permits a comparison to the comparable values for the external membrane system. In general, the biodegradative efficiencies of the submerged membrane system were always lower than those of the external system, although the differences were not significant. This difference in efficiency resulted in estimates of "refractory" wastewater COD for the submerged system that are somewhat higher than those from the external system. For a parameter of this nature, the lowest measured or estimated values are most likely to be closest to the true values. Using this logic, it may be concluded for this wastewater that the COD that is refractory to AnMBR treatment is probably in the range of 70–80 mg COD/L.

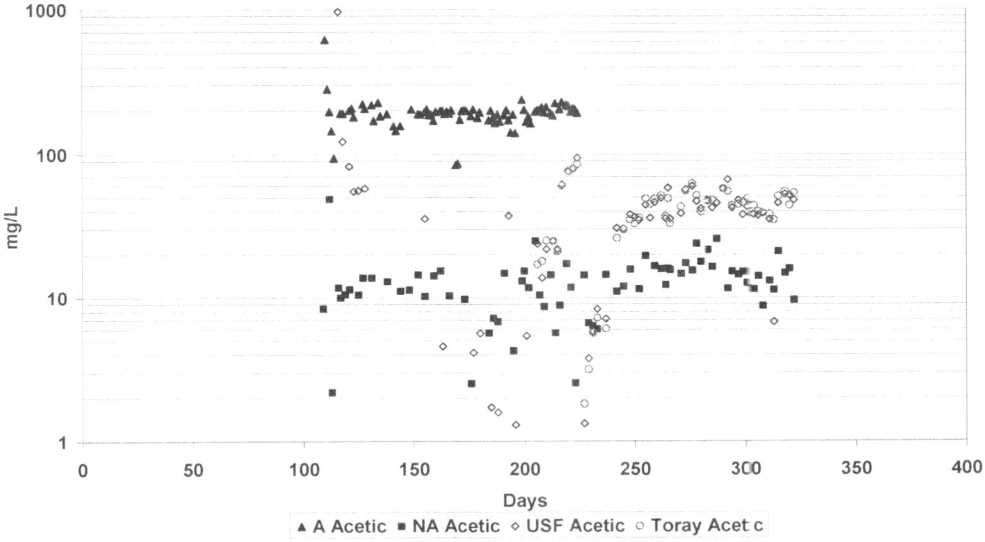

Figure 3-17. Acetic Acid Concentrations in Submerged Membrane AnMBR Influent and Permeates.

Table 3-4. Average Measured Effluent (Permeate) COD, VFA-COD, and "Refractory" COD Concentrations in External and Submerged Membrane AnMBRs During the Periods of Operation with and without Acetate Supplementation.

Parameter	External System	Submerged System
With Acetate Supplementation		
Permeate COD (mg/L)	87	110
Permeate VFA-COD (mg/L)	5	19
Estimated "Refractory" COD (mg/L)	82	93
Without Acetate Supplementation		
Permeate COD (mg/L)	104	142
Permeate VFA-COD (mg/L)	32	58
Estimated "Refractory" COD (mg/L)	72	87

3.2.1.3 Biogas Production

Figure 3-18 summarizes measured and expected total biogas daily production rates for the submerged membrane system. As was discussed above for the external membrane system, measured biogas production was always lower than the expected values and the reason(s) for this are the same as those discussed previously.

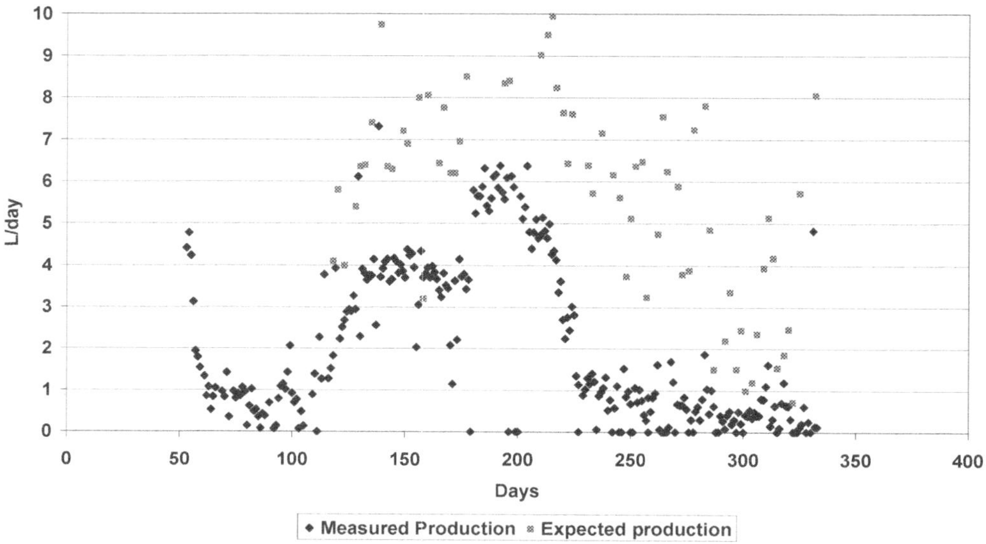

Figure 3-18. Measured and Expected Total Biogas Production Rates in External Membrane AnMBR.

3.2.1.4 Bioreactor SS

The submerged membrane AnMBR bioreactor MLSS concentrations are summarized in Figure 3-19. A steady increase in MLSS concentrations was observed until about day 250, after which the measured concentrations were somewhat erratic, although the SS volatility remained consistently near 80%. As noted previously, it is believed that most of the observed SS accumulation can be attributed to filtration and retention of influent particulate material. In the last three to four months of the study, mixed liquor was unintentionally lost as a result of membrane cleaning events and minor mixed liquor spills resulting from equipment malfunctions.

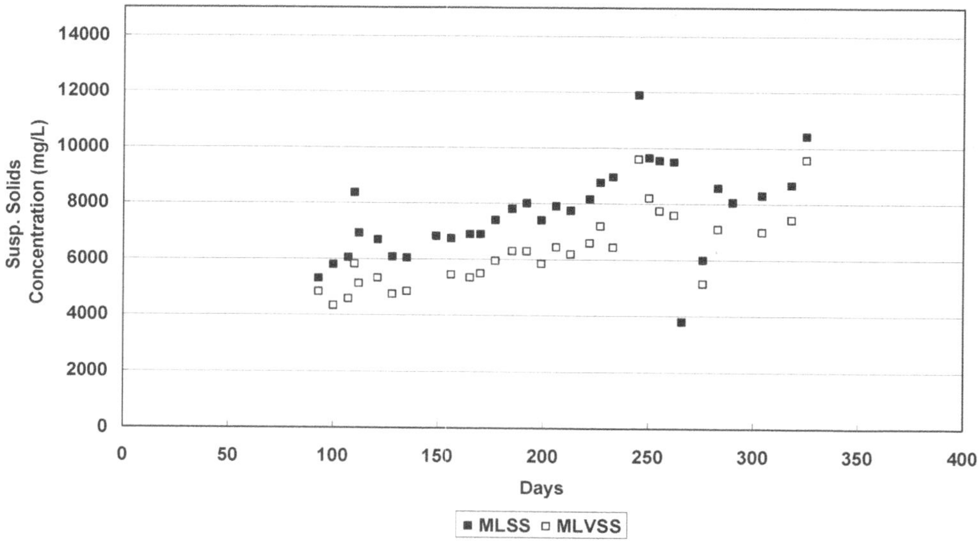

Figure 3-19. Submerged Membrane AnMBR MLSS Concentrations.

3.2.1.5 Bioreactor SS Particle Size Distribution

Figure 3-20 shows the particle size distributions measured during the period of operation between February and May 2005 (days 178 to 262). The SS particle size distribution ranged from approximately 0.2 μm to over 0.2 mm. In contrast to data from the external membrane AnMBR, the percentage of volume of particles in the submerged system was not constant for particles of different sizes. A larger volume percentage was observed for particles in the 10 μm to 0.1 mm range, indicating that overall, there were more larger particles, and fewer smaller particles in the mixed liquor from the submerged membrane AnMBR, compared to that for the external membrane AnMBR.

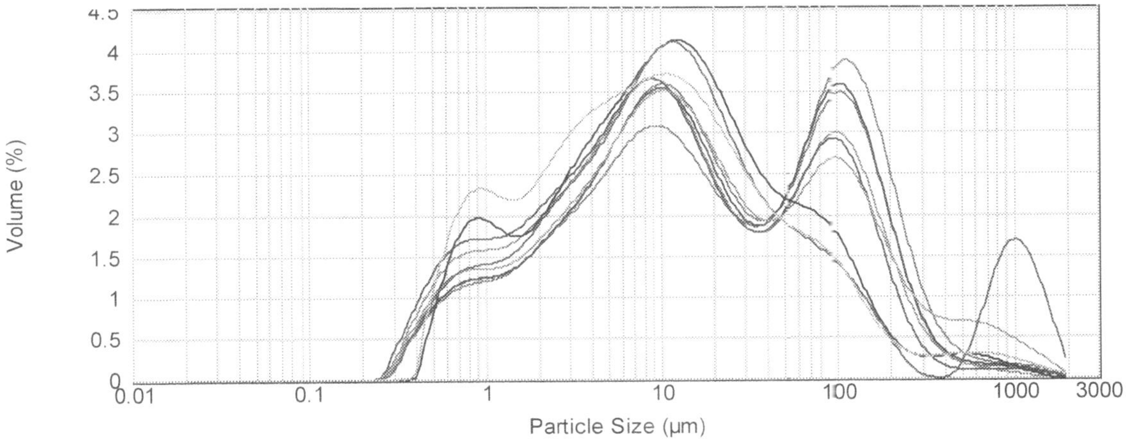

Figure 3-20. Particle Size Distributions of Mixed Liquor from Submerged Membrane System. (Results Presented for Nine Separate Periodic Sampling Events During Phase I)

3.2.1.6 Membrane Flux and TMP

Figure 3-21 and Figure 3-22 show the TMP and flux histories for the Memcor (hollow fiber) and the Toray (flat sheet) membranes. The Memcor TMP increased steadily and the flux declined steadily from about 20 L/m^2·hr at the beginning of the study to values as low as 2.1 L/m^2·hr near the end of the program. The membrane was cleaned a number of times in the last half of the study. However, the resulting TMP decrease and flux increase were short-lived in every case and the rate of TMP increase and flux decline following cleaning became greater after each cleaning event. The Toray TMP also increased and the flux also declined steadily, but the rate of deterioration was lower than that of the Memcor unit. The Toray unit contributed a relatively small fraction of the overall filtration surface area in the submerged membrane AnMBR system, so the operation of the process was almost completely dependent on the flux provided by the Memcor units. Near the end of the study, the total available flux in the submerged unit was insufficient to maintain the intended influent flow rates. Therefore, as noted earlier in Figure 3-12, the nominal bioreactor HRTs in the last 4 months of operation tended to drift from values of approximately 5 hours immediately after membrane cleaning, to about 12 hours before the next cleaning.

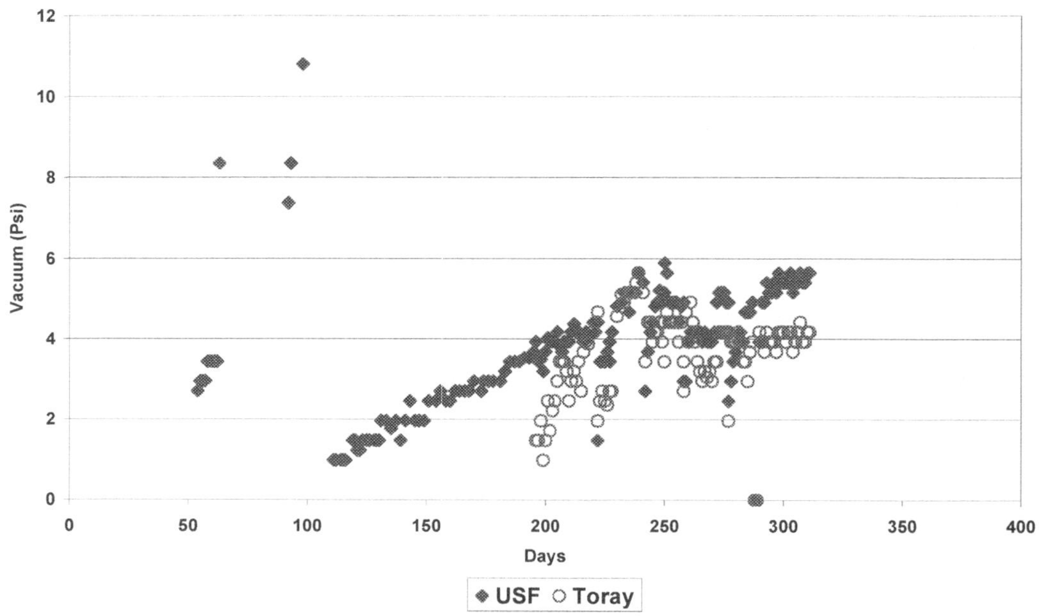

Figure 3-21. Measured TMP of US Filter Memcor and Toray Membranes in the Submerged Membrane AnMBR System.

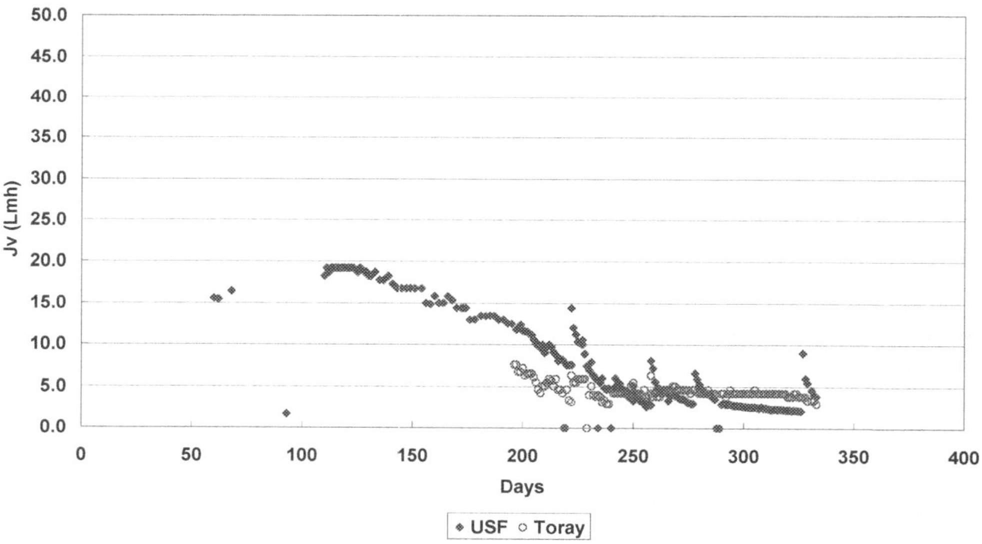

Figure 3-22. Measured Flux of US Filter Memcor and Toray Membranes in the Submerged Membrane AnMBR System.

3.2.2 Phase II

3.2.2.1 Description of Operating Conditions

The target HRT for this period of the project was 8–10 hours and, as shown in Figure 3-23, this HRT was maintained consistently and without difficulty.

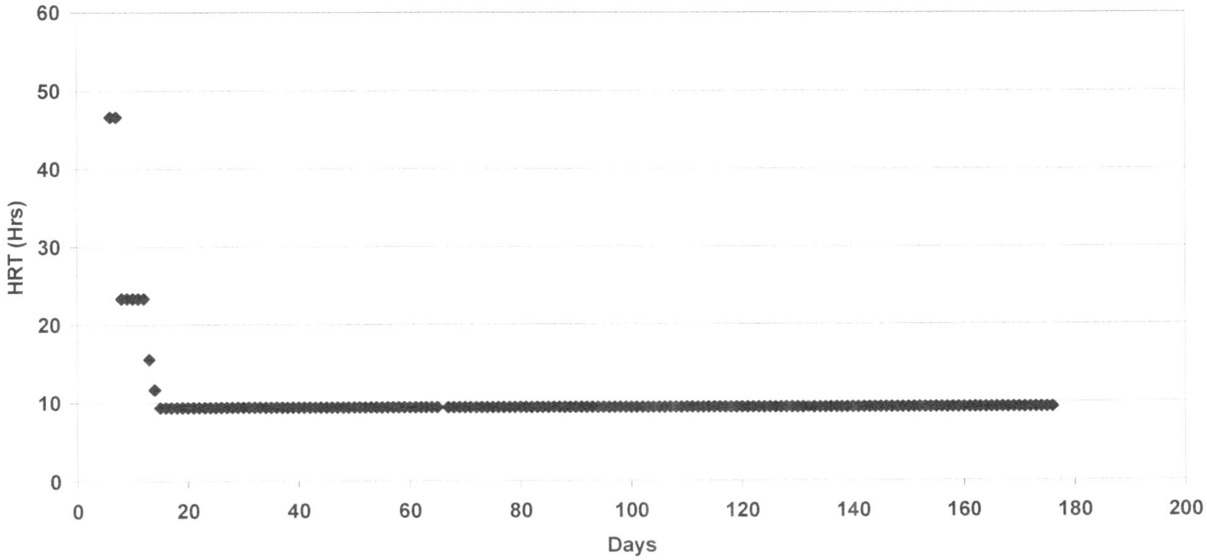

Figure 3-23. HRT of the Submerged MBR.

The OLRs applied during this period of the project are presented in Figure 3-24. They ranged from 1.0 to 1.5 kg COD/m^3·day, largely depending on the wastewater COD concentration. Note that for Phase II, the influent wastewater was supplemented with acetic acid to a concentration of approximately 200 mg/L in the influent to the AnMBR.

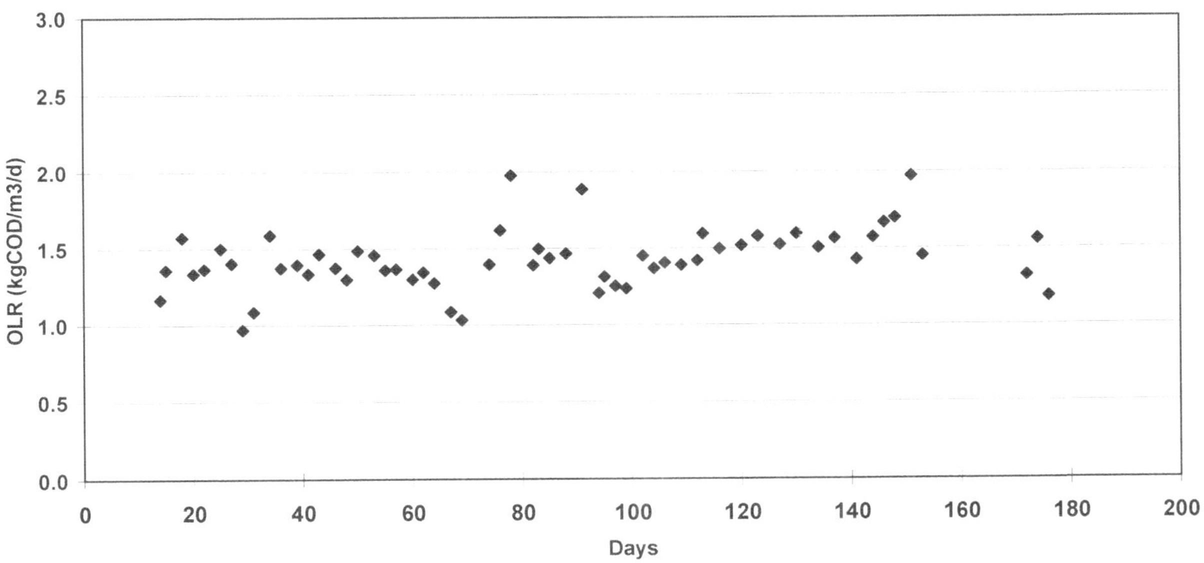

Figure 3-24. OLR of the Submerged MBR.

The pH and the temperature variations in the AnMBR are shown in Figure 3-25 and Figure 3-26, respectively. Since small amounts of sodium bicarbonate were added to the influent wastewater on daily basis (a requirement of another research program), the pH values were consistently higher than 7. The pH value of the influent and effluent were almost same for this period of the study.

The target temperature for this phase of the project was 35°C. The heater used for temperature control was unable to maintain a temperature of 35°C even when the reactor was insulated. Nonetheless, for much of Phase II, the temperature was consistently around 32°C. As discussed in Section 3.2.2.5, new membrane modules were installed in the AnMBR on day 67 of Phase II. The gas sparging mechanism in the new modules was slightly different from that of the old modules. The gas sparging in the new modules appeared to cause less cooling than previously and therefore, it became possible to maintain the temperature at 35°C (Figure 3-26).

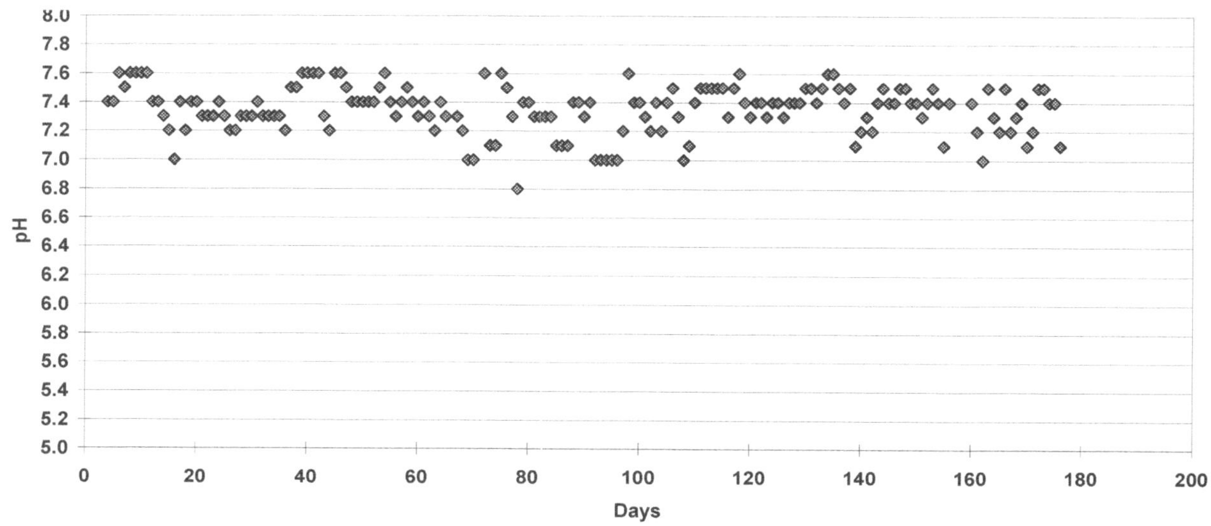

Figure 3-25. Effluent pH of the MBR.

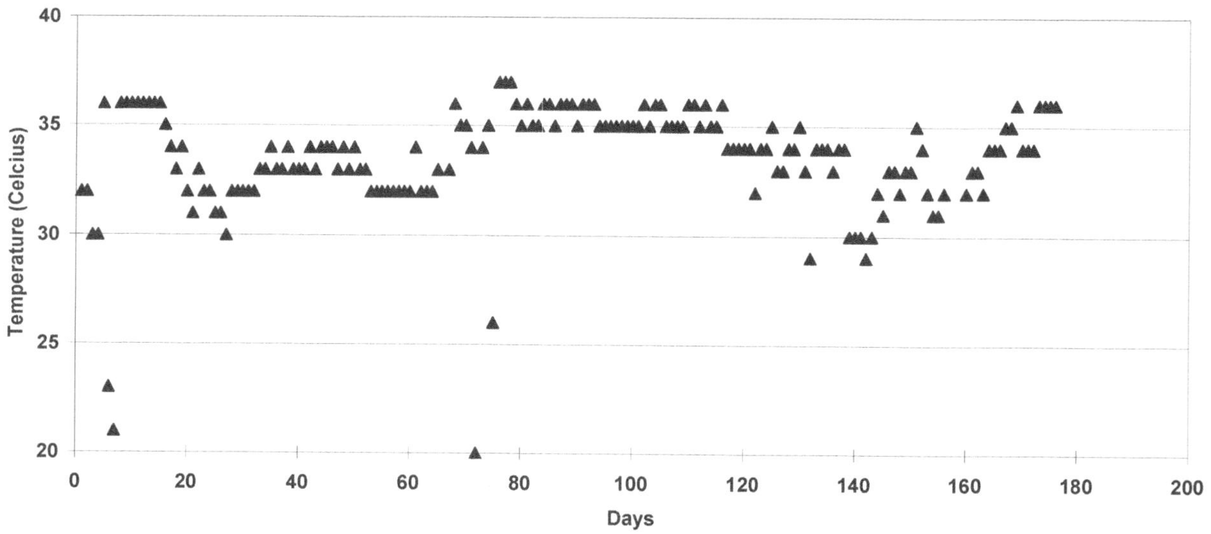

Figure 3-26. Temperature of the Submerged MBR.

3.2.2.2 Treatment Performance

With acetate supplementation, the AnMBR influent contained an average of approximately 470 mg COD/L, of which approximately 350 mg COD/L was soluble. The raw wastewater total COD and soluble COD concentrations were approximately 260 and 130 mg/L, respectively. The measured influent and permeate COD concentrations are presented in Figure 3-27. The effluent COD concentration was consistently below 80 mg/L and the total COD removal efficiency averaged 85%. It is important to note that this removal efficiency applies to an influent supplemented with acetate. When the COD removal efficiency was calculated using raw wastewater total COD (instead of acetate supplemented wastewater total COD), the average removal efficiency was substantially lower—72% (Table 3-5).

As presented in Figure 3-28, the observed VFA removal was essentially complete throughout Phase II, as very low permeate VFA concentrations were measured (6–8 mg/L).

The observed average of 72% COD removal undoubtedly represents the best indicator of the potential COD removal that could be achieved under the anaerobic treatment conditions applied in this study. In contrast to the results obtained in Phase I, the performance data obtained during Phase II were produced while both the bioreactor and the membrane units were performing consistently well. The low concentrations of VFAs in the permeate indicate that the anaerobic biological treatment stage was functioning satisfactorily during this time. Even so, the average observed permeate COD concentration of 73 mg/L suggests that AnMBR treatment alone probably cannot achieve a secondary quality effluent.

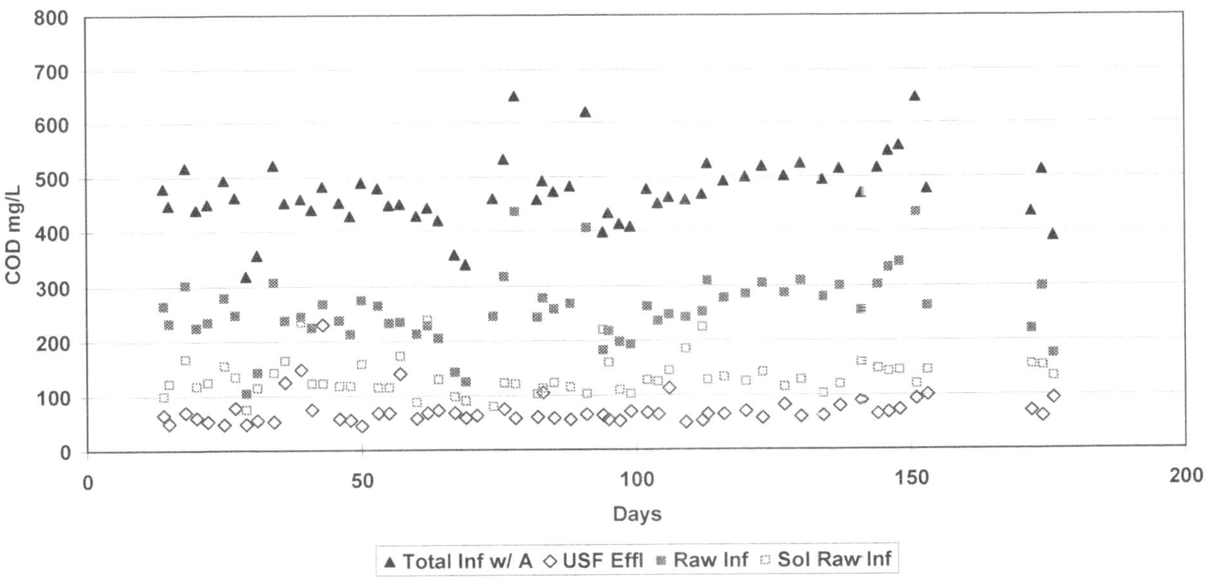

Figure 3-27. Total and Soluble COD Concentrations in Influent and Permeates for the Submerged AnMBR System. (w/A: Influent with Acetate Supplementation)

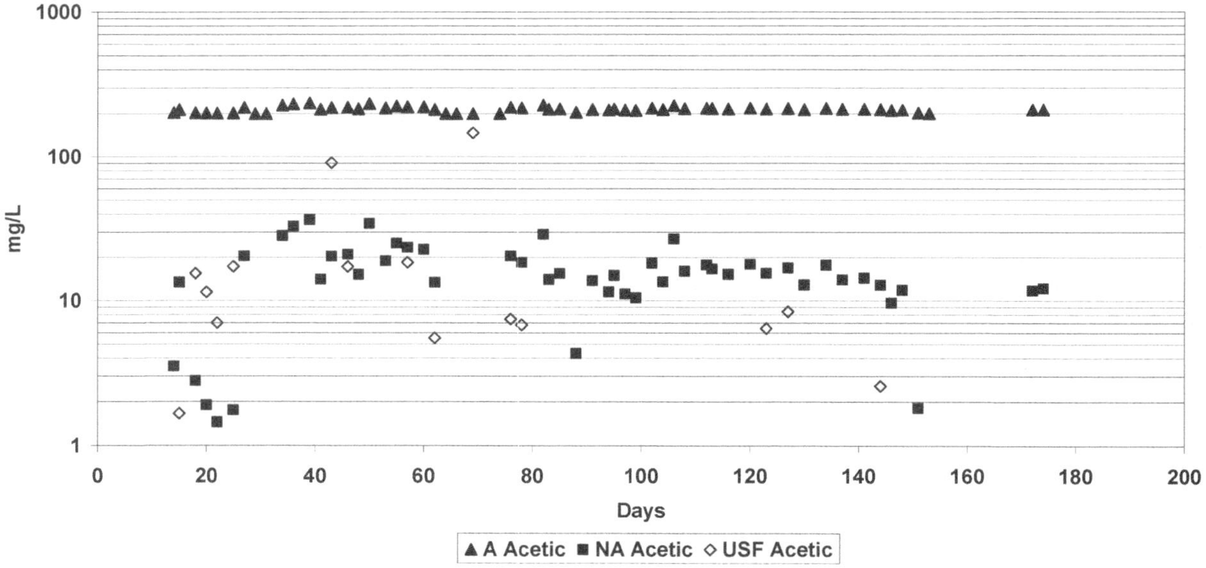

Figure 3-28. Acetic Acid Concentrations in the Submerged Membrane AnMBR Influent and Permeate. (A: Influent with Acetate Supplementation; NA: Influent without Acetate Supplementation)

As previously discussed in Sections 3.1.2 and 3.2.1.2 (Phase I), a substantial amount of the permeate COD was thought to be composed of material that was refractory to anaerobic biodegradation under the conditions applied. The amount of this refractory COD was estimated by assuming that the membrane permeate COD was composed of only two categories of COD: volatile acid COD (i.e., biodegradable COD), and anaerobically nonbiodegradable COD (i.e., "refractory" COD). Table 3-5 indicates that, after subtraction of the VFA-COD from the permeate COD, the remaining 67 mg/L represents the estimated average concentration of refractory COD. This value is somewhat lower than the range of values (72–93 mg/L) that were calculated in a similar fashion using Phase I data (Tables 3-2 and 3-4).

Table 3-5. Average Phase II Measured Wastewater and Permeate COD, VFA-COD, "Refractory" COD, and Residual BOD Concentrations, with Removal Efficiencies.

Parameter	Value
Average Influent COD (wastewater origin) (mg/L)	258
Average Permeate COD (mg/L)	73
Wastewater COD Removal Efficiency (%)	72
Average Permeate VFA-COD (mg/L)	6
Estimated Average "Refractory" COD (mg/L)	67

To provide additional insight into the nature of the residual material in the treated permeate, four sets of raw wastewater and AnMBR permeate samples were taken for analysis of $CBOD_5$ at the end of the study (days 155 to 165). As indicated in Table 3-6, the average $CBOD_5$ concentration in the raw wastewater (without acetate supplementation) was 158 mg/L, while the associated average concentration in the permeate was 56 mg/L. These data result in an estimated $CBOD_5$ removal efficiency of 65%. As noted above for COD, this figure probably represents the optimum $CBOD_5$ removal efficiency that could be expected under the treatment conditions.

Of additional interest is the estimated ratio of 5-day biochemical oxygen demand (BOD_5)/COD in the permeate. The ratio of 0.76 indicates that much of the residual COD that was assumed to be refractory under anaerobic conditions, was in fact, rapidly biodegradable under aerobic conditions. It would be useful to examine the composition of this residual material to determine whether any improvement in anaerobic treatment could somehow be gained, or whether aerobic post-treatment would be compulsory to achieve substantially lower levels of permeate COD and BOD_5.

Table 3-6. Averages of Measurements of BOD_5 in Selected Raw Wastewater and Permeate Samples Taken During Phase II (with Standard Deviations)

Parameter	Value
Average Influent $CBOD_5$ (wastewater origin) (mg/L)	158 ± 18
Average Permeate $CBOD_5$ (mg/L)	56 ± 6
Wastewater $CBOD_5$ Removal Efficiency (%)	65
Estimated Ratio of Permeate $CBOD_5$/COD	0.76

3.2.2.3 Biogas Production

On average, with acetate supplementation, the expected biogas production rate was approximately 8.3 L/d, and the measured biogas production rate was approximately 4.6. The arguments advanced earlier in Section 3.1.3 would also apply here to explain the discrepancy between measured and expected values. From day 67 of Phase II onward, the measured total gas production was periodically very low as shown in Figure 3-29. This is due to the fact that the AnMBR had to be opened several times to service the new membrane modules and for the on-line testing.

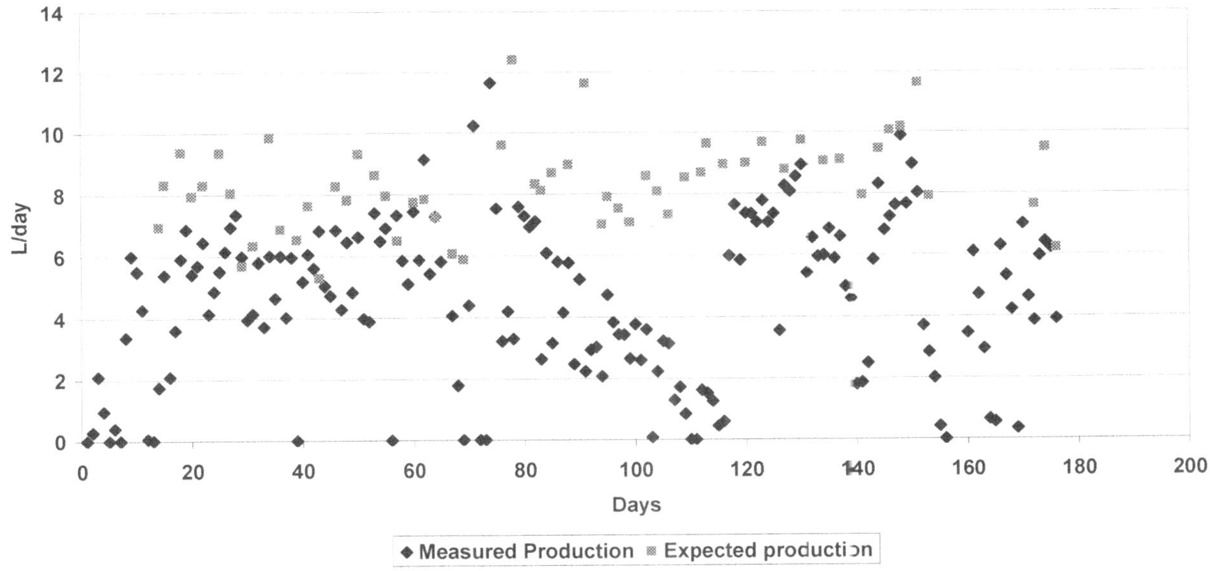

Figure 3-29. Measured and Expected Total Biogas Production Rates in the Submerged Membrane AnMBR.

The composition of the AnMBR off-gas is summarized in Figure 3-30. The irregularity in the percentage of nitrogen and methane results from the opening of the reactor to access the new membrane modules. Every time the reactor was opened, the reactor headspace was purged with nitrogen gas to prevent oxygen from contacting the mixed liquor. In Figure 3-31, the gas composition data have been adjusted by eliminating the N_2 component so that the relative proportions of the CH_4 and CO_2 can be discerned. When the data are expressed in this fashion, it can be seen that the anaerobic off-gas generally contained 80-95% CH_4 and 5-20% CO_2. Figure 3-30 and Figure 3-31 confirm that the proportions of CO_2 and CH_4 produced during Phase II were relatively constant.

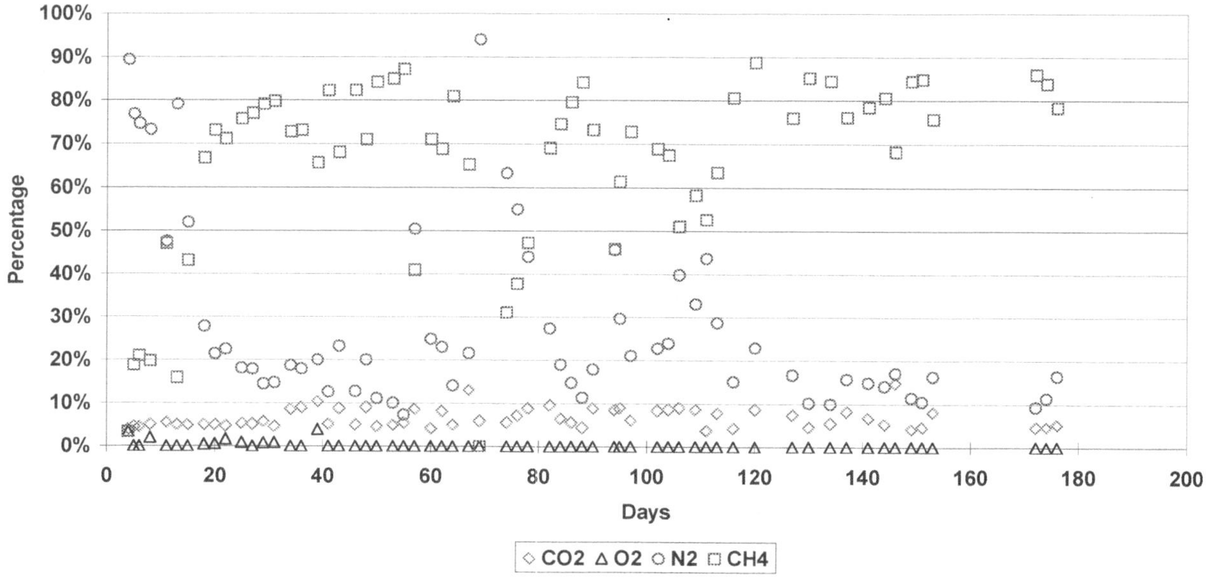

Figure 3-30. Measured Composition of Biogas from the Submerged MBR.

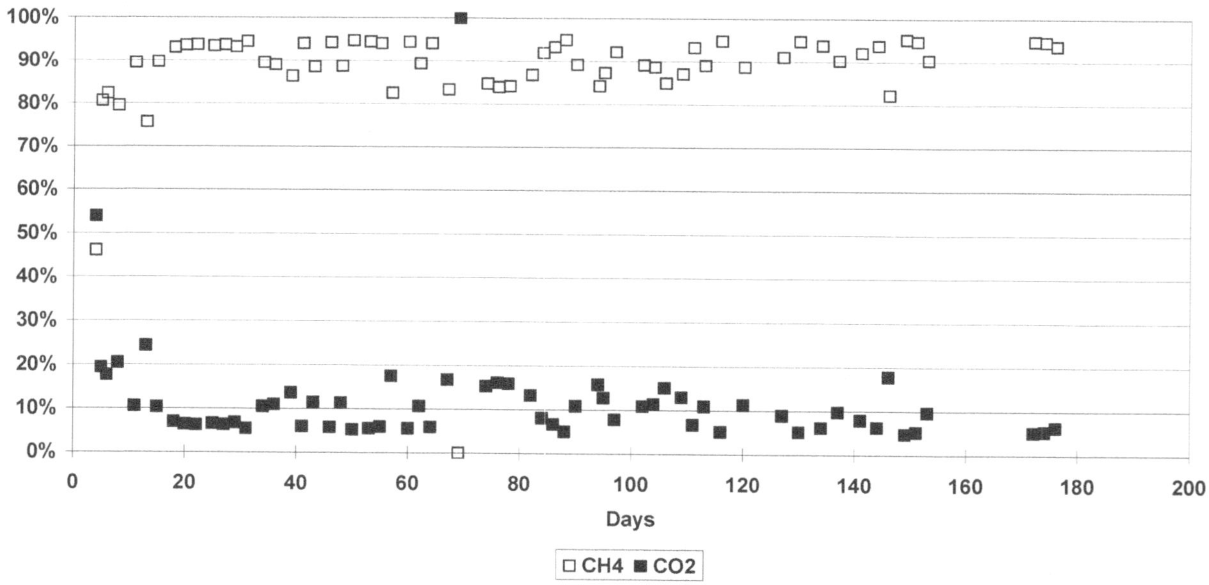

Figure 3-31. Adjusted Composition of CH_4 and CO_2 Only of Biogas from the Submerged MBR.

3.2.2.4 SS Concentration

During the entire Phase II experimental period, no sludge wasting was practiced. As a result, until day 67, the MLSS concentration increased to a value of approximately 14,000 mg/L. However, on day 67, a substantial amount of sludge was lost when the new membranes were installed. As a result, the MLSS concentration dropped to approximately 5000 mg/L. As presented in Figure 3-32, the MLSS concentration subsequently increased to about 10,000 mg/L.

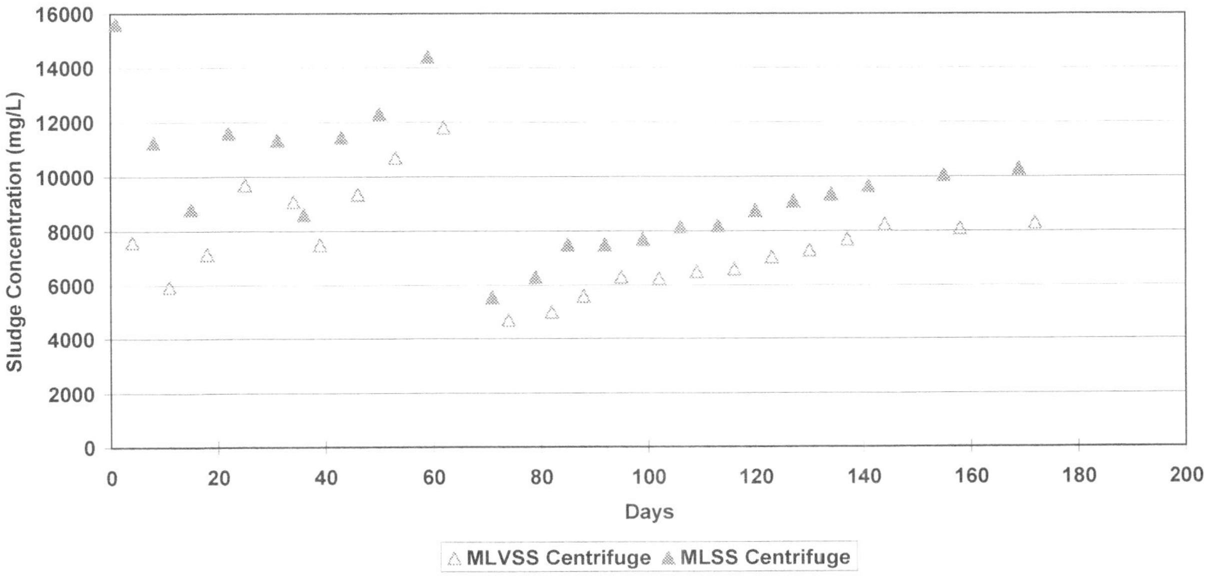

Figure 3-32. Submerged MBR Sludge Concentration.

3.2.2.5 Membrane Flux and TMP

On day 67 of Phase II, new membrane modules were installed in the AnMBR since it was no longer possible to effectively clean the old membrane modules (i.e., following chemical cleaning of the old membrane modules, the TMP would increase relatively rapidly as presented in Figure 3-33). The new membrane modules are essentially identical to the old ones, except for a slightly different gas sparging system.

The new membranes enabled a more consistent flux to be maintained as presented in Figure 3-34. However, the new membranes were not in operation for a period of time that was comparable to that of the old ones, which were used throughout Phase I and the first 67 days of Phase II. In addition, as presented in Section 3.2.1.6, during Phase I, the old membranes were typically operated with a permeate flux in excess of 5 $L/m^2 \cdot h$. Therefore, caution must be taken when comparing the permeate flux that can be maintained when comparing the "new" and "old" membrane modules. Toward the end of the Phase II period, the membrane flux was increased to approximately 12–14 $L/m^2 \cdot h$ for the on-line filtration testing (Figure 3-34), and this resulted in a more rapid increase in TMP, as indicated in Figure 3-33.

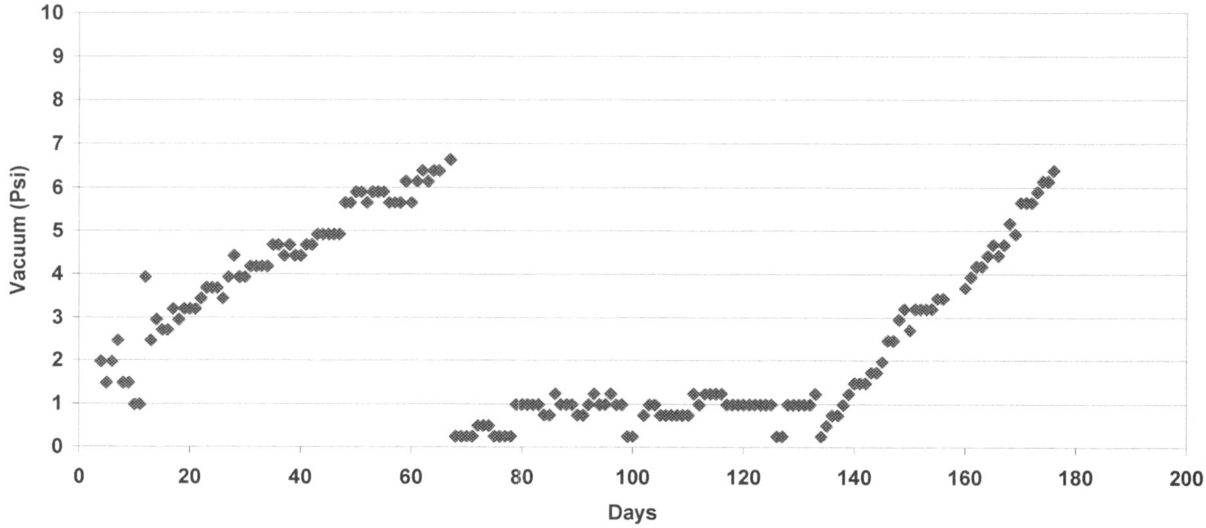

Figure 3-33. TMP of the Submerged MBR.

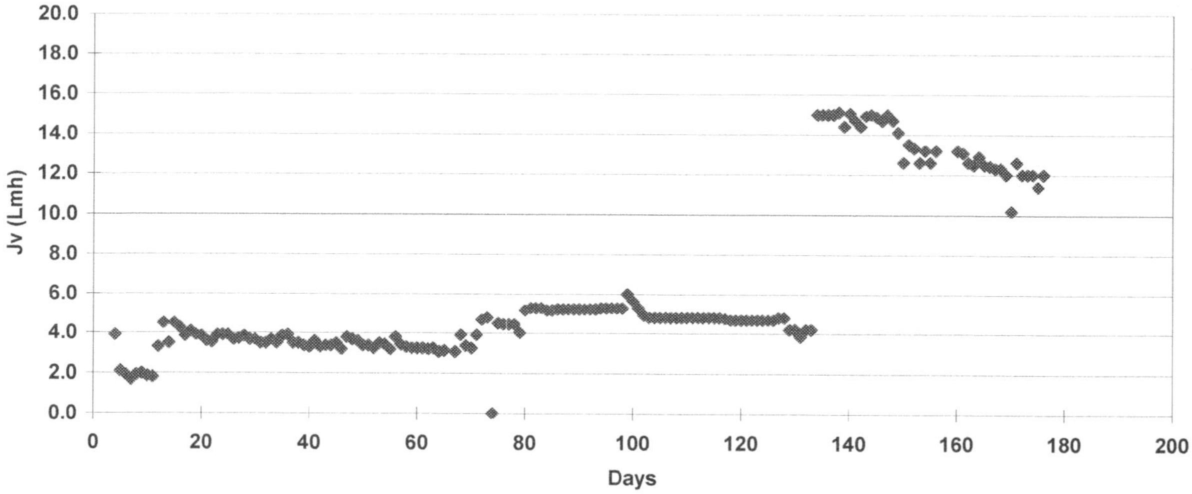

Figure 3-34. Flux of the Submerged MBR.

3.3 Discussion

The premise at the outset of this project was the assumption that the anaerobic treatment process itself did not need to be the focus of the study. Instead, it was decided that the research would focus on a comparative evaluation of the operational performance of a variety of different AnMBR system configurations. Both external and submerged membrane AnMBR configurations were investigated. For the external membrane system, two types of polymeric tubular membranes, and one type of inorganic (i.e., ceramic) tubular membrane were considered. For the submerged membrane system, two types of polymeric membranes, hollow fiber and flat sheet membranes, were considered. Municipal wastewater was chosen as the substrate for the study because it was deemed to be a challenging feed for an AnMBR. For the first part of Phase I and all of Phase II, the influent was supplemented with acetic acid. This approach was taken because of concerns that the limited concentrations of biodegradable organic material in the municipal wastewater would not be sufficient to sustain a consistently robust anaerobic methanogenic environment in the laboratory-scale AnMBRs. Because the study focus was on membrane performance rather than bioreactor performance, it was felt that the addition of limited quantities of acetate would not materially alter the membrane filtration impacts of the wastewater under treatment. All AnMBRs were operated at low-to-typical mesophilic temperatures (25–35°C).

3.3.1 Anaerobic Treatment Performance

The efficiency of the anaerobic treatment observed in the study was dependent on operating conditions. For the first half of Phase I, the systems were operated at temperatures in excess of 30°C and with acetate supplementation of the municipal wastewater feed, both AnMBRs were capable of achieving about 80% total COD removal and about 70% soluble COD removal. It was estimated that the refractory COD remaining in the treated effluent following AnMBR treatment ranged between 70-80 mg/L. This level of anaerobic treatment was achieved in the external membrane AnMBR at HRTs of about 10 hours with OLRs of between 1.0 and 1.5 kg COD/m^3·d. In the submerged membrane AnMBR, similar COD removal efficiencies were observed at HRTs of 101–13 hr and associated OLRs of about 0.8–1.2 kg COD/m^3·d.

In Phase II, the operating conditions utilized resulted in steady, satisfactory performance of the submerged membrane AnMBR. The treatability information collected under these conditions indicated that 72% COD removal and 65% $CBOD_5$ removal could be achieved on average. The measured residual concentrations of COD and $CBOD_5$ in the permeate averaged 72 mg/L and 56 mg/L respectively, far greater than typical secondary effluent levels. These figures likely represent the optimal treatment performance that could be obtained under the operating conditions applied in this study. It was further demonstrated that most of the permeate COD was biodegradable under aerobic conditions, suggesting that post-aerobic polishing of the AnMBR permeate could potentially be used to effectively reduce the effluent COD concentration to low levels.

During the second half of Phase I, both AnMBR systems were operated without acetate supplementation and with increasingly more challenging HRTs and temperatures. Under these more challenging conditions, average total COD removal decreased to 57% in the external membrane systems and 44% in the submerged membrane system. As influent flow rates were increased to operate with HRTs of less than 5 hours in both systems, membrane flux became the controlling factor in system operation and performance.

There are very few literature reports that describe similar research using *real* municipal *wastewater*. The literature review prepared as part of the present project identified only three such studies. Stuckey and others at Imperial College have operated submerged membrane AnMBRs with *synthetic sewage*, reporting effluent carbonaceous BOD_5 values of less than 30 mg/L for operation at 35°C and HRTs as low as 3 hr (Stuckey and Hu, 2003). Pagilla's team at the Illinois Institute of Technology used an external membrane AnMBR to treat municipal primary effluent and achieve COD removal efficiencies that were comparable to those observed in the present study (55–68%), at 32°C and HRTs as low as 12 hr (Baek and Pagilla, 2003;

Pagilla, 2004). The most impressive performance was reported by Wen at Tsinghua University, using a hollow fiber, submerged membrane system with internal fixed film media (Wen et al., 1999). Wen's group also reported effluent $CBOD_5$ values of less than 30 mg/L but at temperatures and HRTs as low as 12°C and 4 hr respectively. A significant difference between the present study and that of Wen et al. is the strength of the influent wastewater. In the present study, the highest influent COD concentrations were about 520 mg/L. In the Wen et al. report, the influent COD was as high as 2,600 mg/L.

The operating conditions imposed in the present study appear to be comparable to those reported in the three studies cited above. Further, at least under the best operating conditions applied in the present study, the observed COD removal efficiencies and effluent qualities appeared to be similar to those reported in the literature. Of the two process configurations operated in the present study, the external membrane AnMBR appeared to produce marginally better effluent quality (lower COD and volatile acids) than the submerged membrane system, under marginally higher loadings.

During the second half of Phase I of the study, membrane cleaning was undertaken repeatedly to maintain the target process flows. With both AnMBR systems, membrane cleaning was disruptive. The isolation or removal of the membrane units always introduced air into the systems and usually also resulted in some loss of mixed liquor. Near the end of the program, when membrane cleaning was very frequent, these disturbances appeared to affect process integrity significantly, particularly in the submerged membrane system.

3.3.2 Membrane Filtration Unit Performance and Behavior

One of the primary objectives of the study was to assess the performance of different commercially available membranes in an AnMBR applied to the treatment of municipal wastewater. In total, five membrane units were used at one time or another in the study.

3.3.2.1 External Membranes

A PCI tubular polymeric membrane was removed from service in the external membrane AnMBR near day 150, after the flux had declined to untenable levels. It is likely that the PCI membrane system failed because it was not possible to maintain a sufficiently high CFV through all five parallel tubes of the membrane unit. Due to its parallel tube configuration, the highest CFV that could be achieved through the PCI membrane system was approximately 0.5 m/s. For the remainder of the study, two Membralox ceramic single tube membranes in series and and four Koch polymeric single tube membranes in series were used with the external membrane MBR. It was possible to achieve a CFV of approximately 2 m/s for both the Membralox and Koch single tube membranes in series. These results are consistent with those of other studies that indicate that the rate of fouling in an AnMBR decreases as the CFV increases (Grethlein, 1978; Elmaleh and Abdelmoumni, 1997; Bourgeous et al., 2001).

3.3.2.2 Submerged Membranes

In the submerged membrane AnMBR, a Memcor hollow fiber unit was used throughout the study and a segment of a Toray flat sheet membrane was installed in the submerged system for the last half of the Phase I program.

The membranes installed in each of the two AnMBR systems were operated in parallel and in each MBR system, the membranes filtered the same mixed liquor. When utilized in this manner, the qualities of the permeates generated by the two membranes in each system were essentially the same. This was observed for both the external and submerged membrane MBR systems. It appeared that permeate quality was not substantially impacted by the membrane used for filtration, but rather, by the operation and performance of the bioreactor portion of the MBR process. Perhaps this outcome is not surprising, as the nominal pore sizes of the membranes used in the study were similar in all cases.

Of the four membrane units studied, the flux characteristics of the Membralox ceramic membrane appeared to be the most robust to the changing AnMBR operating conditions employed. Early in the study, the Membralox unit exhibited a slow deterioration in flux that was not reversible by cleaning. Thereafter, the Membralox flux remained steady (Figure 3-11) at about 110 $L/m^2 \cdot hr$ or greater, for about 5 months of operation at relatively short HRTs and a low temperature. The Membralox flux deteriorated significantly only over the last month of operation, during which time the bioreactor performance was poor. The Koch membrane flux was much lower on average (50 $L/m^2 \cdot hr$) and at the end of the program, the deterioration of the Koch flux began earlier and became more extensive than with the Membralox unit. Again, membrane cleaning was not successful in restoring flux in the Koch unit to any significant extent. Both Kang et al. (2002) and Ghyoot and Verstraete (1997) observed that the extent of fouling for inorganic membranes in an AnMBR was less than that for polymeric membranes. However, Ghyoot and Verstraete (1997) noted that the life cycle cost of ceramic membranes is approximately twice that of polymeric membranes.

As shown in Figure 3-21 and Figure 3-22, both of the submerged membranes exhibited a TMP increase and a flux decline. Although the Memcor unit was cleaned a number of times in the last half of Phase I, the recovered TMP and flux quickly deteriorated to pre-cleaning values, after which the overall slow decline in flux continued. Membrane cleaning in a submerged system is messy and problematic for the biological stage of the process. For membrane cleaning, the bioreactor needs to be opened and the biogas is unavoidably lost and replaced by air. Then the membrane unit needs to be removed from the bioreactor, cleaned, and replaced; all of which introduces considerable stress to the anaerobic biomass. For both the external and submerged AnMBRs, a cleaning sequence that repeatedly alternated between high pH sodium hypochlorite cleaning and citric acid cleaning was needed to effectively recover the membrane permeability. A simple high pH sodium hypochlorite cleaning was not sufficient to effectively recover the membrane permeability. This suggests that the foulant consisted of both organic and inorganic material (Kang et al., 2002). These results were confirmed using SEM-EDX microanalysis as discussed in the next chapter.

The results clearly indicate that the dominant particle sizes in the external membrane system (0.1–0.2 μm) were significantly smaller than those in the submerged system (10–150 μm). It is believed that this major difference in particle size distribution can be attributed to the differences in turbulence and shear between the two AnMBRs. In the submerged membrane AnMBR bioreactor, mixing and membrane fouling control was provided by recirculation of biogas to sparging systems located beneath the Memcor and Toray membrane units. Although the intensity of biogas sparging was fairly high, the resulting shear on the mixed liquor was probably much lower than that of the external membrane system. The operation of the external membranes was driven by TMP and CFV. TMP was controlled by manipulation of a throttling valve downstream of the parallel membrane systems (Figure 2-1). An adequate CFV was required to affect fouling control at the filtration surface of each membrane. The required CFV could only be maintained by running the progressing cavity mixed liquor recirculation pumps at a very high rate (R), relative to the influent flow rate (F). The intensive recirculation pumping and the passage of the recirculation stream through the throttling valve, imparted considerable energy and shear to the external membrane mixed liquor. It seems logical that this added shear would result in substantially smaller particles in the external membrane system, relative to those in the submerged system. It is believed that the impact of the large R/F ratios is exaggerated by the scale of the external membrane AnMBR used in the laboratory studies. If the same diameter external membrane modules were to be used in a larger scale system, the required cross-flow velocities could be maintained at lower R/F ratios. This should have the effect of reducing average shear intensities to which the biomass is exposed and presumably, mixed liquor particle size distributions would be increased. Choo and Lee (1996a, 1998) reported that fine colloids play a critical role in increasing the hydraulic resistance of a foulant layer in an AnMBR. Fine colloids tend to have a lower back-diffusion rate than larger solids. Choo and Lee (1998) suggested that, as a consequence of this lower back-diffusion rate, fine colloids tend to migrate to, and accumulate at, the membrane surface to a greater extent than larger SS. In addition, they suggested that smaller solids tend to form a more compact foulant layer on the membrane surface. Therefore, not only do smaller solids tend to accumulate at the membrane surface, the resulting foulant layer is more compact. Consequently, one would expect that there should be more fouling in the external AnMBR system. However, this was not the case in the present study. It is likely that other factors, such as differences in the amount of shear imposed on external and submerged membranes, also impacted the permeate flux.

As expected, due to the low TMPs, the fluxes in the submerged membrane AnMBR were much lower than those of the external system. It should be noted however, that the TMP affects more than just the driving force for permeation through the membrane. Permeate flux in an AnMBR is governed by different parameters when operated at low or high TMPs. At relatively low TMPs, such as those typically used in submerged membrane systems, permeate flux is governed by TMP. Under such pressure-limited conditions, permeate flux increases linearly with the applied TMP, and permeate flux is not significantly affected by CFV (Beaubien et al., 1996). At relatively high TMPs, such as those typically used in external membrane systems, permeate flux is governed by the mass transfer of material away from the membrane surface. Under mass-transfer-limited conditions, permeate flux in an AnMBR is governed by CFV (i.e., surface shear) and MLSS concentration (Beaubien et al., 1996; Choi et al., 2005). Beaubien et al. (1996) reported a linear increase in permeate flux with an increase in CFV along the membrane surface at high TMPs. Choi et al. (2005) also observed an increase in permeate flux when the CFV was increased in an AnMBR treating swine manure. However, a plateau was reached, above which a further increase in CFV did not result in an increase in permeate flux. These results suggest that, as the CFV increases, mass-transfer limitations can be removed, and the system becomes pressure-limited. The impact of the TMP on the driving force, the different fouling mechanisms that prevail at low and high operating TMPs that are typically present in submerged and external membrane systems, respectively, and the significantly different hydrodynamic conditions present in submerged and external membrane systems, which impact the size of the material in the mixed liquor as discussed above, all combine to make a direct comparison between the magnitude of the permeate flux in submerged and external systems difficult in absolute terms. To simplify the comparison between the external and submerged membrane systems, in the present study, all membrane systems were operated as close as possible to the manufacturer recommended operating conditions in terms of TMPs, sparging intensities, and CFVs.

CHAPTER 4.0

ASSESSMENT OF THE MEMBRANE FILTRATION CHARACTERISTICS IN A SUBMERGED ANMBR USING OFF-LINE AND ON-LINE FILTRATION TESTS

As presented in the previous chapter, substantial differences were observed between the filtration characteristics for external and submerged AnMBRs. Also, the filtration characteristics were observed to be different for a comparison of the AnMBRs and an anaerobic-anoxic-aerobic MBR treating the same wastewater under the auspices of a separate study. Off-line filtration tests were performed during Phase I of the present study to compare the filtration characteristics of an external and a submerged AnMBR, as well as between a submerged AnMBR and a submerged aerobic MBR. The results of the off-line filtration tests performed during Phase I of the study are presented in Section 4.1. Additional on-line filtration tests were performed during Phase II to further explore the differences observed during Phase I between the submerged AnMBR and the submerged anaerobic-anoxic-aerobic MBR. The results of the on-line filtration tests performed during Phase II of the present study are presented in Section 4.2.

4.1 Off-Line Filtration Tests (Phase I)

A series of five off-line tests were performed as part of Phase I of the study (see Section 2.1.4.2). The results from these tests are presented below.

4.1.1 Off-Line Filtration Tests Performed on Mixed Liquor from the Submerged Aerobic MBR

The anaerobic-anoxic-aerobic sludge used in Test 1 was obtained from the UBC pilot scale membrane enhanced biological phosphorus removal process (referred to as the aerobic MBR), which was treating the same municipal wastewater as that used in the AnMBR study. The results for the aerobic mixed liquor served as a benchmark (Figure 4-1) to which anaerobic mixed liquor filtration results could be compared.

New membrane fiber loops were submerged in activated sludge and operated under vacuum with aeration. The resistance increased over the first 30 min and then leveled off (Period I). After 120 min with a constant resistance, the sludge was replaced with DDW and a 5-min membrane relaxation period was applied (Period II). During this period, a significantly lower resistance was observed. After a quick 30-sec back flushing, the resistance decreased again, but only slightly (Period III).

The results of this test indicated that *Rc* contributed most of the overall filtration resistance for the aerobic mixed liquor. The irreversible resistance due to adsorption, *Ra*, only accounted for a very small portion of the total resistance of the foulant layer, *Rs*. Over longer filtration cycles, the contributions of the irreversible resistance (Ra) would be expected to be larger.

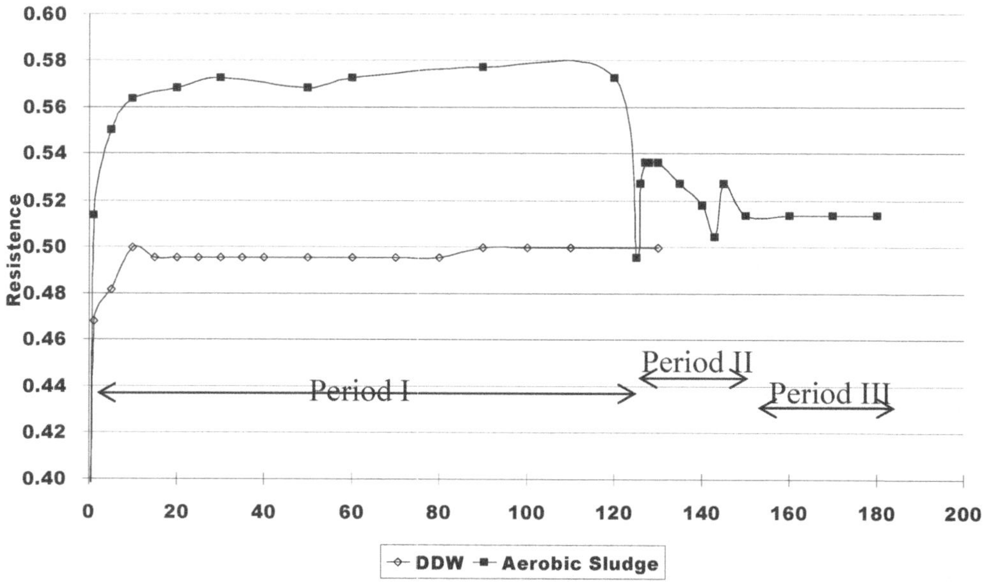

Figure 4-1. Aerobic Sludge Filtration Test Results. (Period I: Filtration of Activated Sludge; Period II: Filtration of DDW; Period III: Filtration of DDW after 30-sec Back-flushing; Resistances expressed in units of m^{-1})

4.1.2 Filtration Testing with Submerged Membrane AnMBR Sludge

Tests 2 and 3 were conducted to investigate the filtration characteristics of the mixed liquor retained in the submerged membrane AnMBR. Test 2 was conducted with undiluted mixed liquor taken from the submerged AnMBR (i.e., with an MLSS concentration of 20000 mg/L), while Test 3 was completed with the same mixed liquor, diluted to a SS concentration that was comparable to that of the aerobic MBR mixed liquor (i.e., with an MLSS concentration of 5,000 mg/L).

The results for the undiluted mixed liquor are shown in Figure 4-2. When filtering the higher concentration mixed liquor, *Rs* increased rapidly in the first 20 min, remained relatively constant for approximately 40 min and then decreased slightly before stabilizing at a lower value (Figure 4-2). However, when filtering the lower concentration mixed liquor (Figure 4-3), *Rs* stabilized after the initial increase and did not decrease thereafter. However, this phenomenon did not occur during the test with the aerobic sludge. This is due perhaps to the greater resistance caused by the anaerobic sludge. Despite identical operating conditions for both the aerobic and anaerobic filtration tests, it was established that the anaerobic sludge developed much higher filtration resistance than the aerobic sludge. The observed resistance caused by the filtration of the anaerobic mixed liquor appeared to be caused mainly by a cake layer. Due to the short term nature of the off-line testing *Ra* and *Rp* accounted for a very small portion of the overall resistance.

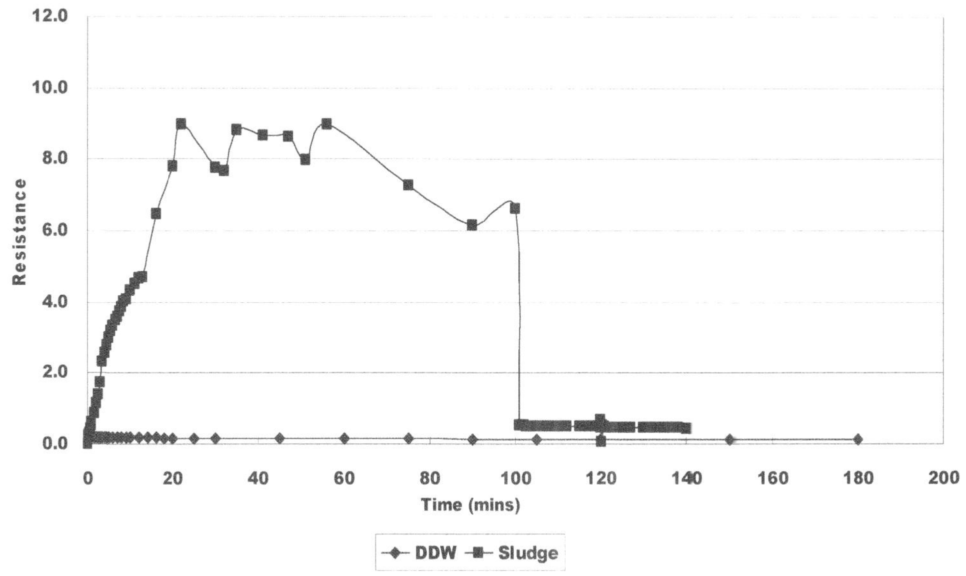

Figure 4-2. Submerged Membrane AnMBR Sludge Filtration Test Results — Test 2 — Undiluted Mixed Liquor. (Period I: Filtration of Aerobic Mixed Liquor; Period II: Filtration of DDW; Period III: Filtration of DDW after 30 sec Back-flushing; Resistances expressed in units of m^{-1})

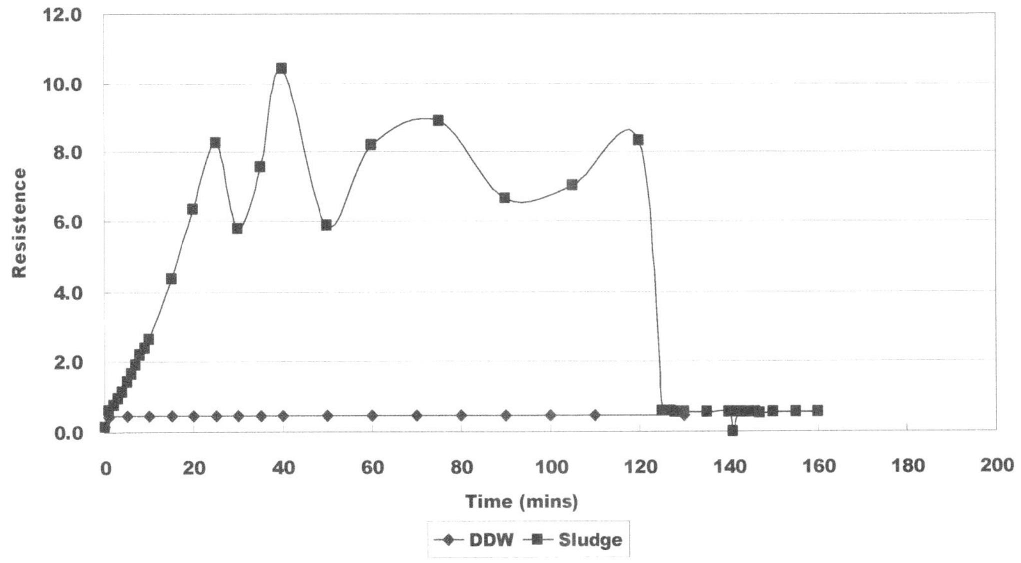

Figure 4-3. Submerged membrane AnMBR sludge filtration test results — Test 3 — Mixed liquor diluted to 5,000 mg/L. (Period I: filtration of anaerobic mixed liquor; Period II: filtration of DDW; Period III: filtration of DDW after 30 sec back-flushing; Resistances expressed in units of m^{-1})

Although the sludge concentration did not play a significant role in the total resistance response, (Figure 4-2 and Figure 4-3), it did however, affect *Ra*. This might be due to the higher colloidal solids concentration available at the higher mixed liquor concentration, which may tend to migrate towards and accumulate at the membrane surface to a greater extent. A higher dissolved-solids concentration in terms of SMPs associated with undiluted mixed liquor may also play a contributing role. Because the resistance due to adsorption is irreversible by hydraulic fouling-reduction measures such as gas sparging and back-flushing, chemical cleaning would be required.

4.1.3 Filtration Testing with External Membrane AnMBR Sludge

Tests 4 and 5 investigated the filtration characteristics of the mixed liquor in the external membrane anaerobic bioreactor. As above, Test 4 was conducted with undiluted mixed liquor while Test 5 was conducted with mixed liquor diluted to 5,000 mg/L. The test results are shown in Figure 4-4 and Figure 4-5.

The filtration results were found to be quite similar to those observed with the submerged AnMBR mixed liquor. The total resistance for the undiluted filtration test (Test 4) increased for the first 20 min and then remained relatively constant. Again, *Rc* seems to have contributed the most to membrane fouling, as shown in Figure 4-4 and Figure 4-5.

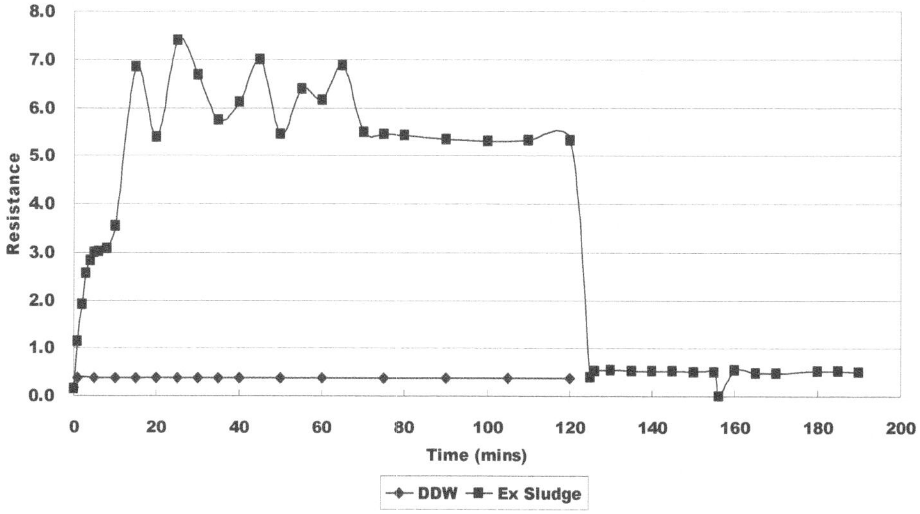

Figure 4-4. External Membrane AnMBR Sludge Filtration Test Results — Test 4 — Undiluted Mixed Liquor. (Period I: Filtration of Anaerobic Mixed Liquor; Period II: Filtration of DDW; Period III: Filtration of DDW after 30 sec Back-flushing; Resistances expressed in units of m^{-1})

For mixed liquors from both the submerged and external systems, *Rs* was attributed primarily to *Rc,* and this resistance was not affected by the mixed liquor SS concentration to a significant extent. In addition, the magnitudes of the overall resistances were relatively similar for both the external and submerged membrane AnMBRs.

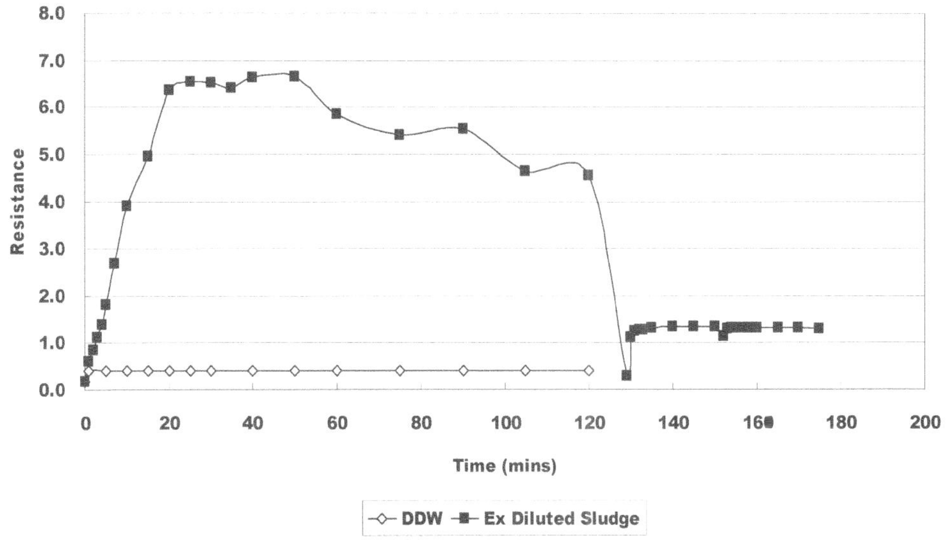

Figure 4-5. External Membrane AnMBR Sludge Filtration Test Results — Test 5 — Mixed Liquor Diluted to 5,000 mg/L. (Period I: Filtration of Anaerobic Mixed Liquor; Period II: Filtration of DDW; Period III: Filtration of DDW after 30 sec Back-flushing; Resistances expressed in units of m^{-1})

The filtration characteristics of the anaerobic sludges were significantly different from those of the aerobic sludge, as determined during short-term off-line filtration tests. It appeared that short-term membrane fouling caused by anaerobic sludge is mainly attributable to external fouling, in which a cake layer played the most prominent role. Gas sparging and membrane relaxation can effectively reduce the external fouling, whereas back-flushing did not appear to have much effect on membrane fouling control.

4.1.4 SEM-EDX Analysis

Sections of membrane fibers from off-line filtration tests using submerged AnMBR mixed liquor were subjected to SEM-EDX microanalysis. The SEM images indicate that the fouled membranes were covered with bacterial colonies (rod-shaped material) and porous material (amorphous material) as presented in Figure 4-6.

Figure 4-6. Typical SEM Image of Fouled Membrane.

The EDX microanalysis results for the clean and fouled membrane are summarized in Table 4-1. A substantial amount of the material detected during the EDX microanalysis of the virgin membrane surface consisted of carbon and fluorine. This was expected since the membrane fibers used for the off-line tests were made from PVDF. The fouled membrane surface however, did not contain a substantial amount of fluorine, suggesting that the fouled membrane surface was extensively covered by the foulant layer. This would certainly be consistent with the SEM images (Figure 4-6). The EDX microanalysis also suggests that the foulant layer consisted of a broad mixture of materials. A number of studies have suggested that struvite ($NH_4MgPO_4 \cdot 6H_2O$) is a significant component of the foulant layer on a membrane in an AnMBR (Choo et al., 2000). However, in the present study, magnesium accounted for only approximately 1% of the foulant material, suggesting that even if struvite was present, it accounted at most for only approximately 10% of the total foulant material (i.e., approximately 10% of the material in struvite is magnesium). As presented in Figure 4-6, the foulant layer appeared to be dominated by a substantial amount of bacterial colonization. However, nitrogen and phosphorus accounted for only approximately 2% and 5% , respectively, of the foulant material, suggesting that bacterial colonies accounted at most for only approximately 17% of the foulant material, assuming that the carbon:nitrogen:phosphorus ratio for biomass is approximately 60:14:5. Considering that carbon was the most abundant material in the foulant layer, these results suggest that the bulk of the foulant material consisted of nonbiologic organic material.

Table 4-1. EDX Analysis of Off-Line Tests.

a) virgin membrane

Material	Concentration
Carbon	75.47 wt%
Oxygen	4.39 wt%
Fluorine	15.18 wt%
Sodium	0.92 wt%
Magnesium	0.75 wt%
Aluminum	2.62 wt%
Nickel	0.67 wt%

b) fouled membrane

Material	Concentration
Carbon	57.33 wt%
Nitrogen	2.11 wt%
Oxygen	17.78 wt%
Fluorine	0.24 wt%
Sodium	0.78 wt%
Magnesium	1.06 wt%
Aluminum	4.36 wt%
Silicon	2.90 wt%
Phosphorus	5.21 wt%
Sulfur	4.04 wt%
Calcium	2.01 wt%
Titanium	0.50 wt%
Iron	1.03 wt%
Zinc	0.64 wt%

Note: Only compounds that were detected using EDX microanalysis are presented.

4.2 On-Line Filtration Tests (Phase II)

To assess the membrane filtration characteristics of mixed liquors under more relevant permeate flux and hydrodynamic conditions, on-line filtration tests (see Section 2.2.2) were conducted during Phase II in the submerged membrane AnMBR and in a pilot scale anaerobic-anoxic-aerobic MBR (referred to as the aerobic MBR) operating in the biological phosphorus removal mode. The results from these tests are presented below.

4.2.1 Evolution in TMP over Time

The evolution of the TMP over time measured during the on-line tests was used to estimate the extent of fouling in the AnMBR and aerobic MBR. An operating permeate flux of 15 L/m^2·hr was selected for the on-line tests so that both the AnMBR and aerobic MBR systems would be subjected to operating conditions that were typical of those applied in full scale applications. However, it was relatively difficult to maintain a constant permeate flux, particularly for the AnMBR (see Section 3.2.2.5). In addition, the initial TMP when operating at 15 L/m^2·hr was slightly different for the AnMBR and the aerobic MBR. To facilitate the comparison between the extent of fouling in the AnMBR and the aerobic MBR, the TMP measurements were adjusted to a constant permeate flux of 15 L/m^2·hr, and normalized with respect to the initial TMP using Equation 4.1. The normalized TMP is presented in Figure 4-7.

$$TMP_{AN} = TMP_{O}\left(\frac{J_S}{J_O}\right)\left(\frac{1}{TMP_{t=0}}\right) \qquad (4.1)$$

where TMP_{AN} is the adjusted and normalized TMP, TMP_O is the operating TMP, J_S is the selected permeate flux (i.e., 15 L/m^2·hr), J_O is the actual operating permeate flux, and $TMP_{t=0}$ is the operating TMP at the start of a filtration cycle (i.e., following membrane cleaning) when filtering mixed liquor.

For the selected operating permeate flux, the TMP_{AN} for the AnMBR increased relatively rapidly. These results were expected since it was concluded in Phase I of the present study that it was not possible to operate a submerged membrane AnMBR with a permeate flux of greater than approximately 5 L/m^2·hr in a sustainable manner due to rapid fouling (see Section 3.2.1.6). However, the TMP_{AN} for the aerobic MBR remained relatively constant. These results were also expected since sustainable operating permeate fluxes for air-sparged submerged hollow fiber aerobic MBRs typically range from 15 to 30 L/m^2·hr.

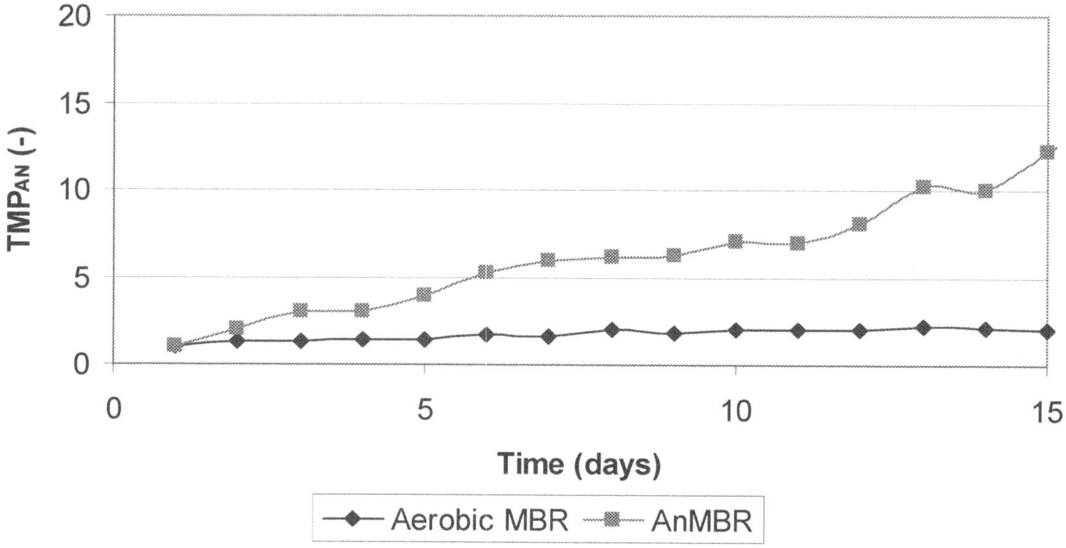

Figure 4-7. Evolution of TMP over Time During the On-Line Filtration Tests.

4.2.2 SMP Analysis of AnMBR and Aerobic MBR Mixed Liquors

The SMP contents of the mixed liquors from both the AnMBR and the aerobic MBR are presented in Table 4-2. The concentrations of carbohydrates in the mixed liquors for both the AnMBR and the aerobic MBR were similar. The concentrations of soluble proteins in the mixed liquor were observed to be slightly higher for the AnMBR than for the aerobic MBR; however, the difference could not be concluded to be statistically significant. The concentrations of humic substances in the mixed liquor for the AnMBR were significantly higher than for the aerobic MBR. It should be noted that during the on-line tests, the MLSS concentration in the AnMBR ranged from approximately 8,000 to 10,000 mg/L, while the MLSS in the aerobic MBR was approximately 6,000 mg/L.

Table 4-2. SMP Analysis of AnMBR and Aerobic MBR Mixed Liquor.

	Carbohydrates (μg/L)		Proteins (μg/L)		Humic Substances (μg/L)	
	Test 1	Test 2	Test 1	Test 2	Test 1	Test 2
AnMBR	42±23	52±15	32±5	21±6	61±5	57±3
Aerobic MBR	29±5	48±17	14±2	18±4	6±1	20±19

Note: Tests 1 and 2 were performed on days 145 and 160, respectively, of Phase II.
±: Standard deviation of analysis done in triplicate

4.2.3 SEM-EDX Analysis

Sections of membrane fibers from the on-line filtration tests were subjected to SEM-EDX microanalysis. As presented in Figures 4-9 and 4-10, the surface of the fouled AnMBR and aerobic MBR membranes appeared to be different. For comparison purposes, a typical SEM image of a clean membrane is presented in Figure 4-8. The fouled surface of the membrane in the AnMBR was covered by a relatively dense foulant layer of amorphous material (Figure 4-9). There was no discernable bacterial colonization of the surface of the membrane in the AnMBR. However, the surface of the membrane in the aerobic MBR was covered by a much more porous amorphous structure containing bacterial colonies (rod-shaped material) (Figure 4-10).

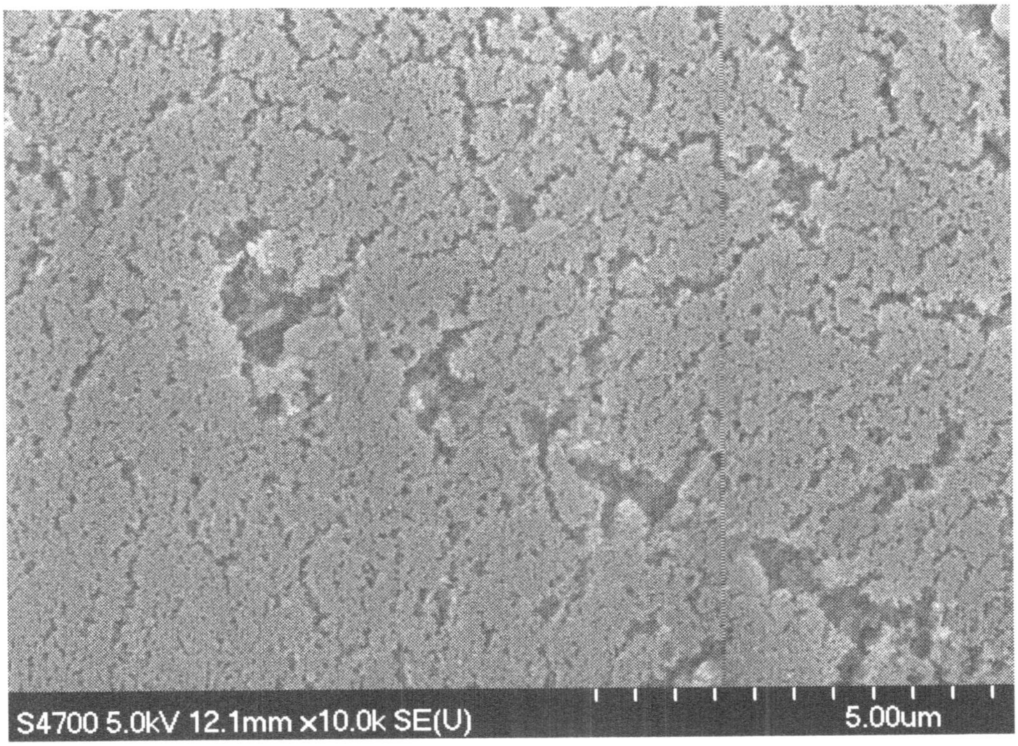

Figure 4-8. Typical SEM Image of Clean Membrane.

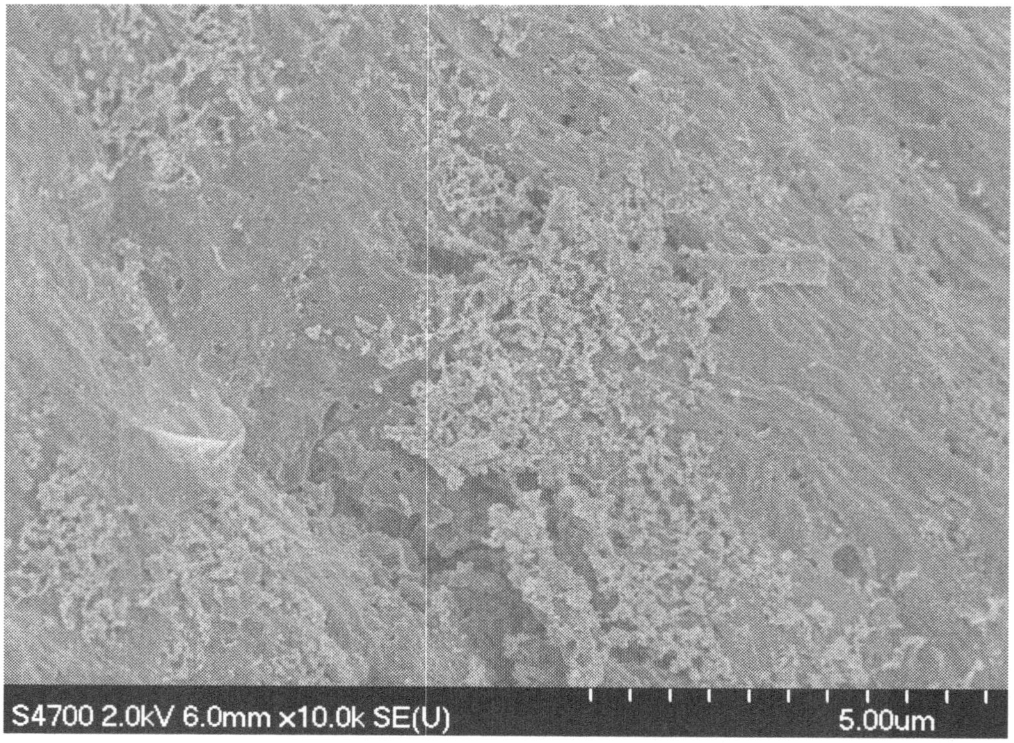

4-9. Typical SEM Image of Fouled Membrane from AnMBR.

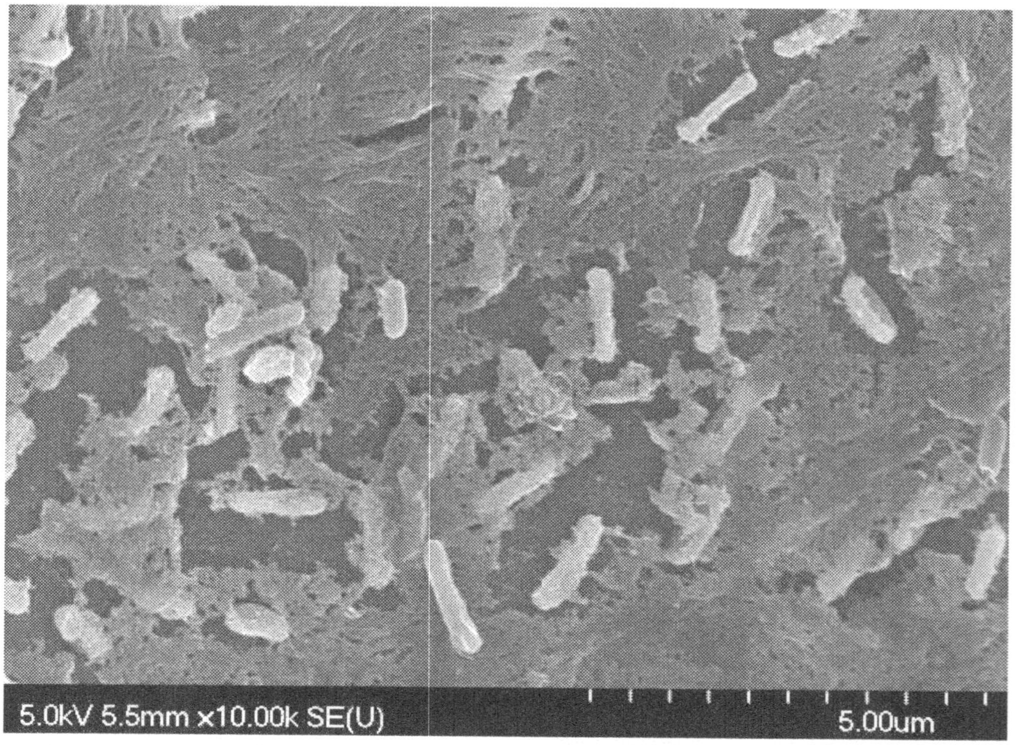

Figure 4-10. Typical SEM Image of Fouled Membrane from Aerobic MBR.

The EDX microanalysis results for the clean and fouled membranes used during the on-line tests are summarized in Table 4-3. It should be noted that the substantially different results from the EDX microanalysis of the clean membranes used for the off-line (Table 4-1) and on-line (Table 4-3) tests were expected since the membrane fibers used for these tests were provided by different manufacturers (see Section 2.1.4.2).

Table 4-3. Typical EDX analysis

a) clean membrane

Material	Concentration (wt %)
Carbon	45.33±0.65
Oxygen	8.12±0.40
Fluorine	43.75±1.06

b) fouled AnMBR membrane

Material	Concentration (wt %)
Carbon	46.65±1.16
Oxygen	5.94±1.55
Fluorine	45.80±2.51

c) fouled aerobic MBR membrane

Material	Concentration (wt %)
Carbon	45.68±2.90
Oxygen	3.42±1.35
Fluorine	49.37±1.83
Nitrogen	1.81±3.14

Notes: ± Corresponds to the standard deviation of the measurements made
Only compounds that were detected using EDX microanalysis are presented

As presented in Table 4-3, the amount of fluorine measured was similar for the clean and fouled membrane surfaces, suggesting that neither of the fouled membrane surfaces was completely covered by a surface foulant layer. The on-line tests were performed using a relatively high gas-sparging intensity, which is thought to generate relatively high shear stresses on the membrane surface. Such high surface shear stresses reduce the extent of surface fouling on membranes (Elmaleh and Abdelmoumni, 1997; 1998).

As presented in Figure 4-10, the surface foulant layer for the aerobic AnMBR appeared to contain a substantial number of embedded microbial cells. However, as presented in Table 4-3, nitrogen accounted for less than 2% of the surface foulant material, suggesting that bacterial colonies accounted at most 15% of the surface foulant material.

4.2.4 Chemical Analysis of Extracted Foulant Material

The TOC, TP, TKN, and SMP contents of the extracted foulant material for both the AnMBR and the aerobic MBR are presented in Tables 4-4 and 4-5.

The TOC concentrations of the acidic and basic extracted foulants for both the AnMBR and the aerobic MBR were relatively similar. The TOC concentrations from the basic extractions were slightly higher than those from the acidic extractions. These results were expected since basic extraction is typically more effective for removing organic foulants from a fouled membrane. The TKN concentrations of the acidic and basic extracted foulants for both the AnMBR and the aerobic MBR were also similar. In addition, the TKN concentrations of the acidic and basic extracted foulants were relatively low. However, the TP concentrations of the extracted foulants were significantly higher for the AnMBR than for the aerobic MBR. In addition, the TP concentrations of the basic extracted foulants were substantially higher than those of the acidic extracted foulants.

Table 4-4. TOC, TP, and TKN Analysis of Extracted Foulant Material.

	TOC (μg/m)		TP (μg/m)		TKN (μg/m)	
	Test 1	Test 2	Test 1	Test 2	Test 1	Test 2
Acidic Extraction						
AnMBR	3,249±244	1,872±162	267±89	121±37	n.d.	n.d.
Aerobic MBR	1,922±353	2,425±617	21±5	26±10	10±8	12±7
Basic Extraction						
AnMBR	5,580±612	2,720±681	1,881±846	898±405	22±2	6±2
Aerobic MBR	4,690±853	3,438±216	5±4	12±6	16±19	24±29

Note: Tests 1 and 2 were performed on days 136 and 156, respectively, of Phase II.
±: Standard deviation of analysis done in triplicate
n.d.: Below detection limit

The carbohydrate concentrations of the acidic and basic extracted foulants for both the AnMBR and the aerobic MBR were similar (Table 4-5). However, significantly higher protein concentrations were measured in the basic extracted foulants for the AnMBR than for the aerobic MBR. (Note: the concentrations of proteins in the acidic extracted foulants were below detection limits.) On the other hand, significantly higher humic substance concentrations were measured in the basic extracted foulants for the aerobic MBR than for the AnMBR. (Note: the concentrations of humic substances in the acidic extracted foulants were below detection limits).

Table 4-5. SMP Analysis of Extracted Foulant Material.

	Carbohydrates (μg/m)		Proteins (μg/m)		Humic Substances (μg/m)	
	Test 1	Test 2	Test 1	Test 2	Test 1	Test 2
Acidic Extraction						
AnMBR	439±161	1,040±485	n.d.	n.d.	n.d.	n.d.
Aerobic MBR	631±102	1,417±583	n.d.	n.d.	n.d.	n.d.
Basic Extraction						
AnMBR	1,438±330	1,364±882	n.a.	1,338±387	n.a.	94±35
Aerobic MBR	1,550±825	1,579±829	n.a.	223±45	n.a.	559±27

Note: Tests 1 and 2 were performed on days 136 and 156, respectively, of Phase II.
±: Standard deviation of analysis done in triplicate
n.d.: Below detection limit
n.a.: Results not available due to contamination of samples during analysis

4.3 Discussion

Off-line filtration tests were performed during Phase I of the study to compare the filtration characteristics of an external membrane and a submerged membrane AnMBR, as well as those of a submerged membrane AnMBR and a submerged membrane pilot scale aerobic MBR. The off-line filtration tests confirmed that the filtration resistances of the anaerobic mixed liquors (external and submerged) were an order of magnitude greater than those of the aerobic MBR mixed liquor produced during treatment of the same wastewater. However, the filtration resistances of both anaerobic mixed liquors were similar.

SEM-EDX microanalysis revealed that the foulant layer that accumulated on the membrane surface during the off-line filtration tests performed on mixed liquor from the submerged AnMBR consisted mainly of non-biological organic material. Results from the EDX analysis suggest that struvite accounted for less than 10% of the total foulant material, and bacterial colonies accounted for less than 18% of the total foulant material present on the membrane surface. These results are consistent with those reported by Choo et al. (2000), who indicated that the cake layer that forms on organic membranes in an AnMBR consists of both biological/organic solids and inorganic precipitates and that the principal inorganic constituent of the cake layer is struvite.

The experimental conditions (i.e., permeate flux and hydrodynamic conditions) used during the off-line filtration tests performed during Phase I of the present study, were significantly different from the conditions applied in the bench-scale AnMBRs. Both the operating permeate flux and the system hydrodynamics (i.e., sparging intensity and system configuration), which greatly affect mass transfer and, therefore, fouling at the membrane surface, were different in the off-line tests and the bench-scale AnMBRs. In addition, the off-line tests only lasted a few hours, while the membranes in bench-scale AnMBRs were operated for a number of months. To assess the membrane filtration characteristics of mixed liquors under more commercially relevant permeate flux and hydrodynamic conditions, on-line filtration tests were conducted during Phase II of the present study.

The difference in the fouling propensities of a submerged AnMBR and a submerged aerobic MBR was assessed by monitoring the TMP of each over time during the on-line tests. The AnMBR membrane fouled much more rapidly than did the aerobic MBR membrane. As previously discussed, the results from Phase I of the present study suggested that it was not possible to sustain a permeate flux greater than approximately 5 L/m^2·hr in a submerged AnMBR. In contrast, it was possible to sustain a permeate flux of 15 L/m^2·hr in a submerged aerobic MBR. Based on discussions with the supplier (US Filter) of the membranes used during Phase II, it was recognized that an operating permeate flux in the order of 15 to 30 L/ m^2·hr can typically be maintained for full-scale sparged submerged hollow fiber aerobic MBRs treating municipal wastewaters.

For aerobic MBRs, the extent of fouling has been extensively documented to be related to the concentration of SMPs in mixed liquor (Chang and Lee, 1998; Lee et al., 2001a; Wisniewski and Grasmick, 1998). In the present study, the concentrations of soluble carbohydrates in the mixed liquors for both the AnMBR and the aerobic MBR were observed to be similar. The concentrations of soluble proteins in the mixed liquor for the AnMBR were observed to be slightly higher than those for the aerobic MBR, however the difference was not statistically significant. The concentrations of soluble humic substances in the mixed liquor for the AnMBR were noted to be significantly higher than those for the aerobic MBR. These results suggest that it may be the higher concentrations of soluble humic substances in the mixed liquor for an AnMBR that may be at least partially responsible for the greater fouling propensity of an AnMBR mixed liquor relative to that of an aerobic MBR. However, higher concentrations of humic substances were measured in the foulant material extracted from the membranes in the aerobic MBR than in that from the membranes in the AnMBR. On the other hand, higher concentrations of proteins were measured in the foulant material extracted from the membranes in the AnMBR than in that from the membranes in the aerobic MBR. Therefore, the results indicate that although the concentrations of soluble proteins in the mixed liquor for the AnMBR were only slightly higher than those for the aerobic MBR, these proteins accumulate to a greater extent on the membranes in an AnMBR than on the membranes in an aerobic MBR. However, the overall concentrations of TOC measured in the foulant material extracted from the membranes for both the AnMBR and the aerobic MBR were comparable, indicating that the overall amount of organic foulant that accumulated on the membranes for both systems were similar.

High concentrations of TP were observed in the foulant material extracted from the membranes in the AnMBR. The relatively high ratio of TP to TOC in the extract suggests that the phosphorus associated with the extracted foulants was likely not associated with the organic material. Choo and Lee (1996a) reported that the principal inorganic foulant in AnMBRs is struvite. Assuming that the phosphorus extracted from the fouled AnMBR membrane was present as struvite and that the phosphorus content of struvite is 11%, the concentration of struvite in the foulant material extracted from the membrane in the AnMBR was approximately twice that of the TOC. However, based on these same assumptions, the concentration of struvite in the foulant material extracted from the membrane in the aerobic MBR was approximately $1/25^{th}$ that of the TOC. Therefore, the results indicate that struvite also may have contributed to the greater fouling propensity in the AnMBR when compared to that of an aerobic MBR.

These results are consistent with the cleaning regimen that was needed to effectively clean fouled AnMBR membranes. As presented in Section 2.1.3, a cleaning sequence that repeatedly alternated between sodium hypochlorite cleaning (at pH 10.5) and citric acid cleaning (at pH 2) was needed to effectively recover the membrane permeability. Lee et al. (2001c) reported that it was not possible to recover permeate flux through an organic membrane in an AnMBR by cleaning the membrane with a caustic solution alone. Caustic solutions are considered to be effective for removing organic/biological foulants from a membrane surface, while acidic solutions are considered to be effective for removing inorganic foulants (Lee et al., 2001c). However, a number of studies have reported that it is possible to consistently recover permeate flux through an organic membrane in an AnMBR by cleaning the membrane with a caustic solution following an acidic cleaning (Lee et al., 2001c). Because the fouling of organic membranes may be attributed to organic material and struvite, the results suggest that removal of deposited struvite governs the recovery of permeate flux.

Considering that phosphorus was not detected by the EDX microanalysis of the fouled membranes for the AnMBR, and that phosphorus was present at relatively high concentrations in the extracted foulants for the AnMBR, it seems that that phosphorus (i.e., struvite) was predominantly associated with internal membrane fouling.

CHAPTER 5.0

CONCLUSIONS AND SUGGESTIONS FOR FUTURE WORK

5.1 Conclusions

5.1.1 Anaerobic Treatment Performance

- Treated effluent quality did not appear to be affected significantly by the type of membrane chosen for an AnMBR application; rather, effluent quality was most impacted by the operation and performance of the anaerobic bioreactor component of the system.

- The potential for COD removal in an AnMBR is best judged from the data collected during the period of stable performance that was associated with Phase II of the study. When the influent was supplemented with acetate, the average total COD removal achieved in the submerged membrane AnMBR was about 80%. Measurements of the COD concentration of the raw wastewater permitted the estimation of the removal efficiency for COD that originated from the wastewater itself. The data available indicated that, on average, AnMBR treatment resulted in 72% removal of the raw wastewater COD, leaving a residual COD concentration of about 73 mg/L in the permeate.

- The average permeate COD concentration of about 73 mg/L observed during Phase II appeared to be refractory to anaerobic biodegradation under the treatment conditions applied.

- A limited amount of BOD_5 data indicated that a submerged membrane AnMBR could achieve about 65% BOD_5 removal, leaving a permeate BOD_5 concentration that averaged 56 mg/L.

- The estimated BOD_5/COD ratio of 0.76 for the AnMBR permeate indicated that much of the organic matter that was refractory to anaerobic treatment, was in fact, aerobically biodegradable. This suggests that aerobic post-treatment may be needed to achieve a secondary quality effluent in terms of BOD_5 or COD.

- The operating conditions and anaerobic treatment efficiencies achieved in the study were comparable to those reported in the very small number of similar previous studies that were cited in the analysis of the relevant compiled literature (WERF Project 02-CTS-4 Task 1. *Phase I Report: Compilation/Review of Existing Literature*).

- Measured biogas production rates were always lower, and in some cases substantially lower, than the expected values. This was attributed to the fact that when low-strength wastewaters are treated anaerobically, a substantial portion of the CO_2 and the CH_4 generated will exit the process dissolved in the treated effluent. In addition, it was recognized that some of the retained particulate COD that was included in the removal efficiency calculation may not be biodegradable and fermentable under the conditions applied.

5.1.2 Membrane Filtration Unit Performance

- Membrane flux became the limiting process attribute for the AnMBR.

- The permeate flux that could be maintained in the external membrane AnMBR (50 to 100 $L/m^2 \cdot hr$) was higher than that which could be maintained in the submerged membrane system (5 $L/m^2 \cdot hr$). However, the external membrane AnMBR was operated at a much higher TMP of (40 psi or 275 kPa) than the submerged system (1 psi or 6.9 kPa).

- Membrane flux maintenance appeared to be superior in the ceramic membrane relative to that in the polymeric membrane. The permeate flux that could be maintained for the polymeric membranes was approximately 50 $L/m^2 \cdot hr$, while that which could be maintained for the ceramic membranes was approximately 100 $L/m^2 \cdot hr$ (at a TMP of 40 psi or 275 kPa).

- Based on discussions with the PSC and a review of relevant literature, it was concluded that the high capital and operating costs associated with external inorganic membranes outweighed the lower fouling tendency of these systems.

- Irreversible fouling was the major contributor to flux decline in both AnMBR systems and this fouling could not be sustained by chemical cleaning. Eventually, irreversible fouling was observed for all of the membranes used in the present study.

- The measured resistance of anaerobic mixed liquor to membrane filtration was much greater than that of mixed liquor from a parallel aerobic MBR treating the same wastewater.

- The submerged membrane AnMBR fouled more readily than the parallel submerged membrane aerobic MBR. The maximum permeate flux that could be sustained in an AnMBR was approximately 5 $L/m^2 \cdot hr$, while the permeate flux that could be sustained in the aerobic MBR was at least 15 $L/m^2 \cdot hr$ (at a TMP of approximately 1 psi or 6.9 kPa).

- Although the concentrations of soluble proteins in the mixed liquor from the submerged membrane AnMBR were only slightly higher than those from the parallel aerobic MBR, these proteins accumulated to a greater extent on the membranes in the AnMBR than on the membranes in the aerobic MBR. However, the overall concentrations of TOC measured in the foulant material extracted from the membranes from both the AnMBR and the aerobic MBR were similar, indicating that the overall amounts of organic foulants that accumulated on the membranes for both systems were similar.

- Struvite may be at least partly responsible for the greater fouling propensity of an AnMBR when compared to that of an aerobic MBR.

- The results suggest that removal of struvite during membrane cleaning governs the recovery of permeate flux.

- Struvite (i.e., phosphorus) was predominantly associated with internal membrane fouling.

5.2 Suggestions for Future Work

- The treatment performance achieved in the present study did not materially exceed that of the few similar reports available in the literature, and thus, does not point to the possibility of achieving secondary quality effluent in an economical manner using AnMBR technology alone. There is likely merit in assessing the efficacy of an anaerobic system, such as an anaerobic moving bed biofilm reactor, to remove the bulk of the organic material from a wastewater, coupled to an aerobic MBR to polish the effluent from the anaerobic bioreactor.

References

An analysis of the relevant compiled literature and other information, on the topic of anaerobic MBR wastewater treatment, was completed and published as a separate WERF Report (*WERF Project 02-CTS-4 Task 1. Phase 1 Report: Compilation/Review of Existing Literature*). All the references cited in the present document, as well as those cited in the *WERF Project 02-CTS-4 Task 1. Phase 1 Report: Compilation/Review of Existing Literature* are presented below.

Anderson, G.K., Saw, C.B., and Fernandes, M.I.A.P. (1986). "Application of porous membranes for biomass retention in biological wastewater treatment processes," *Process Biochemistry,* December, p. 174.

Anderson, G.K., Kasapgil, B., and Ince, O. (1996). "Microbial kinetics of a membrane anaerobic reactor," *Environmental Technology,* Vol. 17, p. 449.

Angenent, L.T., Banik, G.C., and Sung, S. (2000). "Psychrophilic anaerobic pretreatment of low-strength wastewater using the anaerobic migrating blanket reactor," *Proceedings of the WEF 73rd Annual Conference & Exposition,* Anaheim, CA, October.

APHA, AWWA, and WEF. (1999. *Standard Methods for the Examination of Water and Wastewater*. (20) American Public Health Association (APHA), American Water Works Association (AWWA), Water Environment Federation (WEF), Washington, D.C.

Arros, S., Torrijos, M., Gesan-Guiziou, G., Lafforgue, C., Daufin, G., and Merin, U. (2001). "Is the anaerobic membrane bioreactor a valuable process for purification of end-of-pipe dairy wastewater with reduced sludge production?" *International Conference: Membrane Technology for Wastewater Reclamation and Reuse*, edited by G. Oron and A. Bick, p. 98, Tel Aviv, Israel, September.

Arros-Alileche, S., Gesan-Guiziou, G., Lafforgue, C., Daufin, G., and Merin, U. (2002). "What's the membrane role in an anaerobic membrane bioreactor? Simulation," Proceedings of the Membranes in Drinking and Industrial Water Production Conference, Mulheim/Ruhr, Germany, September.

Baek, S.H. and Pagilla, K.R. (2003). "Anaerobic membrane bioreactor for dilute municipal wastewater treatment," Proceedings of the WEF 49 Annual Conference & Exposition, Los Angeles, CA, October.

Bailey, A.D., Hansford, G.S., and Dold, P.L. (1994). "The enhancement of upflow anaerobic sludge bed reactor performance using crossflow microfiltration," *Water Research,* Vol. 28, No. 2, p. 291.

Barker, D.J. and Stuckey, D.C. (1999). "A review of soluble microbial products (SMP) in wastewater treatment systems," *Water Research*, Vol. 33, No. 144, p. 3063.

Barker, D.J., Salvi, S.M.L., Langenhoff, A.A.M., and Stuckey, D.C. (2000). "Soluble microbial products in ABR treating low-strength wastewater," *Journal of Environmental Engineering*, Vol. 126, No. 3, p. 239.

Barker, D.J. and Stuckey, D.C. (2001). "Modeling of soluble microbial products in anaerobic digestion: the effect of feed strength and composition," *Water Environment Research*, Vol. 73, No. 2, p. 173.

Barnes, S., Dalhoff, R., Keller, J., Wilderer, P., and Kendall, L. (2003). "Investigation of membrane processes for the removal of volatile fatty acids," *Water Science & Technology*, Vol. 47, No. 12, p. 191.

Beaubien, A., Baty, M., Jeannot, F., Francoeur, E., and Manem, J. (1996). "Design and operation of anaerobic membrane bioreactors: development of a filtration testing strategy," *Journal of Membrane Science,* Vol. 109, No 2, p. 173.

Bérubé, P.R. and Hall, E.R. (1999). "Effects of kraft evaporator condensate matrix on methanol removal in a high temperature membrane bioreactor," *Water Science & Technology,* Vol. 40, No. 11-12, p. 327.

Bérubé, P.R., Ronteltap, M., and Hall, E.R. (2003). "Impact of recycling rate on the specific substrate utilization coefficient and the observed growth yield in a membrane bioreactor," submitted for publication.

Bilstad, T., Madland, M., Espedal, E., and Hanssen, P.H. (1992). "Membrane separation of raw and anaerobically digested pig manure," *Water Science & Technology,* Vol. 25, No. 10, p. 19.

Bodik, I., Kratochvil, K., Herdova, B., Tapia, G., and Gasparikova, E. (2002). "Municipal wastewater treatment in the anaerobic-aerobic baffled reactor at ambient temperature," *Water Science & Technology,* Vol. 46, No. 8, p. 127.

Bonkoski, W.A. (2003). Personal communication to P.M. Sutton, Vice President, Zenon Environmental, December.

Botha, G.R., Sanderson, R.D., and Buckley, C.A. (1992). "Brief historical review of membrane development and membrane applications in wastewater treatment in Southern Africa," *Water Science & Technology*, Vol. 25, p. 10.

Bourgeous, K.N., Darby, J.L., and Tchobanoglous, G. (2001). "Ultrafiltration of wastewater: effects of particles, mode of operation, and backwash effectiveness," *Water Research,* Vol. 35. p.77.

Brockman, M. and Seyfried, C.F. (1996). "Sludge activity and cross-flow microfiltration – a non-beneficial relationship," *Water Science & Technology*, Vol. 34, No. 9, p. 205.

Brockman, M. and Seyfried, C.F. (1997). "Sludge activity under the conditions of cross-flow microfiltration," *Water Science & Technology*, Vol. 35, No. 10, p. 173.

Budd, W.E. and Okey R.W. (1969). U.S. patent 3,472,765.

Buswell, A.M., Boruff, C.S., and Wiesman (1932). "Anaerobic stabilization of milk waste," *Industrial and Engineering Chemistry*, Vol. 24, p. 1423.

Cadi, Z., Huyard, H., Manem, J., and Moletta, R. (1994). "Anaerobic digestion of a synthetic wastewater containing starch by a membrane reactor," *Environmental Technology - London*, Vol. 15, No. 11, p. 1029.

Chang, I., Le Clech, P., Jefferson, B., and Judd, S. (2002). "Membrane fouling in membrane bioreactors for wastewater treatment." *Journal of Environmental Engineering,* Vol. 128, p. 1018.

Chang, I-S. and Lee, C-H. (1998). "Membrane filtration characteristics in membrane-coupled activated sludge system — the effect of physiological states of activated sludge on membrane fouling," *Delsalination*, Vol. 120, p. 221.

Cheryan, M. and Mehaia, M.A. (1986). "Membrane bioreactors," *Chemtech,* November, p. 676.

Choate, W.T., Houldsworth, D., and Butler G.A. (1983). "Membrane enhanced anaerobic digesters," Proceedings 37th Industrial Waste Conference, Purdue University, West Lafayette, Indiana, Ann Arbor, MI: Ann Arbor Science, p. 661.

Choi, H., Zhang, K., Dionysiou, D.D., Oerther, D.B., and Sorial, G.A. (2005). "Influence of cross-flow velocity on membrane performance during filtration of biological suspension." *Journal of Membrane Science,* Vol. 248, p. 189.

Choo, K-H. and Lee, C-H. (1996a). "Membrane fouling mechanisms in the membrane-coupled anaerobic bioreactor," *Water Research*, Vol. 30, No. 8, p. 1,771.

Choo, K.-H. and Lee, C-H. (1996b). "Effect of anaerobic digestion broth composition on membrane permeability," *Water Science & Technology,* Vol. 34, No. 9, p. 173.

Choo, K-H. and Lee, C-H. (1998). "Hydrodynamic behavior of anaerobic biosolids during crossflow filtration in the membrane anaerobic bioreactor," *Water Research,* Vol. 32, No. 11, p. 3387.

Choo, K-H., Kang, I.J., Yoon, S.H., Park, H., Kim, J.H., Adlya, S., and Lee, C-H. (2000). "Approaches to membrane fouling control in anaerobic membrane bioreactors," *Water Science & Technology*, Vol. 41, No. 10-11, p. 363.

Chung, Y-C., Jung, J-Y., Ahn, D-H., and Kim, D-H. (1998). "Development of two phase anaerobic reactor with membrane separation system," *Journal of Environmental Sciences and Health, Part A: Toxic/Hazard Substances & Environmental Engineering,* Vol. A33, No. 2, p. 249.

DeFrance, L. and Jaffrin, M. (1999). "Comparison between filtrations at fixed transmembrane pressure and fixed permeate flux: application to a membrane bioreactor used for wastewater treatment," *Journal of Membrane Science,* Vol. 152, p. 203.

Delanghe, B., Nakamura, F., Myoga, H., Magara, Y., and Guibal, E. (1994). "Drinking water denitrification in a membrane bioreactor," *Water Science & Technology,* Vol. 30, No. 6, p. 157.

Ejner, P. (2003). Personal communication to P.M. Sutton, President, Bioscan A/S, October.

Elmaleh, S. and Abdelmoumni, L. (1997). "Cross-flow filtration of an anaerobic methanogenic suspension," *Journal of Membrane Science,* Vol. 131, No. 1-2, p. 261.

Elmaleh, S. and Abdelmoumni, L. (1998). "Experimental test to evaluate performance of an anaerobic reactor provided with an external membrane unit," *Water Science & Technology*, Vol. 38, No. 8-9, p. 385.

Elmitwalli, T.A., Oahn, K.L.T., Zeeman, G., and Lettinga, G. (2002). "Treatment of domestic sewage in a two-step anaerobic filter/anaerobic hybrid system at low temperature," *Water Research*, Vol. 36, No. 9, p. 2,225.

Fakhru'l-Razi, A. (1993). "Performance study of a new membrane anaerobic digester configuration," *Proceedings of the International Conference on Environmental Management, Geo-Water and Engineering Aspects,* p. 115.

Fakhru'l-Razi, A. (1994). "Ultrafiltration membrane separation for anaerobic wastewater treatment," *Water Science & Technology,* Vol. 30, No. 12, p 321.

Fakhru'l-Razi, A. (1995). "Steady states study in a membrane anaerobic system (MAS) for wastewater treatment," *Proceedings of the WEF 68th Annual Conference & Exposition,* Vol. 3, p. 669.

Fakhru'l-Razi, A., and Noor, M.J.M.M. (1999). "Treatment of palm oil mill effluent (POME) with the Membrane Anaerobic System (MAS)," *Water Science & Technology*, Vol. 39, No. 10–11, p. 159.

Fane A.G., Fell C.J.D., and Suzuki, A. (1983). "The effect of pH and ionic environment on the ultrafiltration of protein solutions with retentive membranes," *Journal of Membrane Science*, Vol. 16, p. 195.

Ferras, E., Minier, M., and Goma, G. (1986). "Acetonobutylic fermentation: improvement of performances by coupling continuous fermentation and ultrafiltration," *Biotechnology and Bioengineering*, Vol. 28, p. 523.

Flaschel, E., Raetz, E., and Renken, A. (1986). "Shear deactivation of enzymes in membrane reactors," *Proceedings of the International Conference on Bioreactor Dynamics,* Cambridge, UK.

Foresti, E. (2002). "Anaerobic treatment of domestic sewage; established technologies and perspectives," *Water Science & Technology*, Vol. 45, No. 10, p. 181.

Frølund, B., Palmgren, R., Keiding, K., and Nielsen, P. H. (1996). Extraction of extracellular polymers from activated sludge using a cation exchange resin. *Water Research*, 30(8), 1749-1758.

Fuchs, W., Binder, H., Mavrias, G., and Braun, R. (2003). "Anaerobic treatment of wastewater with high organic content using a stirred tank reactor coupled with a membrane filtration unit," *Water Research,* Vol. 37, No. 4, p. 902.

Fukuma, M., Takesada, K., and Yasunishi, A. (1993). "Two-phase anaerobic treatment of wastewater containing cellulose using membrane module in acidogenic phase," *Kagaku Kogaku Ronbunshu,* Vol. 19, No. 5, p. 936 (in Chinese).

Ghyoot, W.R. and Verstraete, W.H. (1997). "Coupling membrane filtration to anaerobic primary sludge digestion," *Environmental Technology* (England), Vol. 18, No. 6, p. 569.

Giorno, L., Chojnacka, K., Donato, L., and Drioli, E. (2002). "Study of a cell-recycle membrane fermentor for the production of lactic acid by Lactobacillus bulgaricus," *Industrial and Engineering Chemistry Research*, Vol. 41, No. 3, p. 433.

Grethlein, H.E. (1978). "Anaerobic digestion and membrane separation of domestic wastewater," *Journal Water Pollution Control Federation,* Vol. 50, p. 754.

Hall, E.R. (1992). "Anaerobic treatment of wastewaters in suspended growth and fixed film processes," *Design of Anaerobic Processes for the Treatment of Industrial and Municipal Wastes,* edited by J.F. Malina Jr. and F.G. Pohland. Lancaster, PA: Technomic Publishing Co. Ltd., p. 41.

Hall, E.R., Onysko, K.A., and Parker, W.J. (1995). "Enhancement of bleached Kraft organochlorine removal by coupling membrane filtration and anaerobic treatment," *Environmental Technology*, Vol. 16, p. 115.

Harada, H., Momonoi, K., Yamazaki, S., and Takizawa, S. (1994). "Application of anaerobic–UF membrane reactor for treatment of a wastewater containing high strength particulate organics," *Water Science & Technology*, Vol. 30, No. 12, p. 307.

He, Y., Li, C., Wu, Z. and Gu, G. (1999). "Study on molecular weight cut-off of the anaerobic ultrafiltration membrane bioreactor," *Zhongguo Jishui Paishui/China Water & Wastewater,* Vol. 15, No. 9, p. 10.

Hernandez, A.E., Belalcazar, L.C., Rodriguez, M.S., and Giraldo, E. (2002). "Retention of granular sludge at high hydraulic loading rates in an anaerobic membrane bioreactor with immersed filtration," *Water Science & Technology,* Vol. 45, No. 10, p. 169.

Hickey, R.F. (2004). Personal communication to P.M. Sutton, Senior Vice President, Ecovation, February.

Hogetsu, A., Ishikawa, T., Yoshikawa, M., Tanabe, T., Yudate, S., and Sawada, J. (1992). "High rate anaerobic digestion of wool scouring wastewater in a digester combined with membrane filter," *Water Science & Technology,* Vol. 25, No. 7, p 341.

Imasaka, T. Kanekuni, N. So, H., and Yoshino, H. (1989). "Cross-flow of methane fermentation broth by ceramic membrane," *Journal of Fermentation and Bioengineering*, Vol. 68, p. 200.

Ince, O., Anderson, G.K., and Kasapgil, B. (1993). "Improvement in performance of an anaerobic contact digester using a crossflow ultrafiltration membrane techniques for brewery wastewater treatment," *Proceedings of the 48th Industrial Waste Conference,* Purdue University, West Lafayette, IN. Chelsea, MI: Lewis Publishers, Vol. 48, p. 551.

Ince, O., Anderson, G.K., and Kasapgil, B. (1995). "Effect of changes in composition of methanogenic species on performance of a membrane anaerobic reactor system treating brewery wastewater," *Environmental Technology*, Vol. 16, No. 10, p. 901.

Ince, O., Anderson, G.K., and Kasapgil, B. (1997). "Composition of the microbial population in a membrane anaerobic reactor system during start-up," *Water Research*, Vol. 31, No. 1, p 1.

Ince, B., Ince, O., Sallis, P.J., and Anderson, G.K. (1998). "Experimental determination of the inert soluble COD fraction of a brewery wastewater under anaerobic conditions," *Environmental Technology*, Vol. 19, p. 437.

Ince, B.K., Ince, O., Sallis, P.J., and Anderson, G.K. (2000). "Inert COD production in a membrane anaerobic reactor treating brewery wastewater," *Water Research*, Vol. 34, No. 16, p. 3943.

Jahn, A. and Nielsen, P.H. (1995). Extraction of extracellular polymeric substances (EPS) from biofilms using a cation exchange resin, "*Water Science and Technology*, Vol. 32, No. 8, p.157.

Jewell, W.J., Switzenbaum, M.S., and Morris, J.W. (1981). "Municipal wastewater treatment with the anaerobic attached microbial film expanded bed process," *Journal Water Pollution Control Federation*, Vol. 53, No. 4, p.482.

Jones, R.M. and Hall, E.R. (1986). "Pilot-scale study of high-rate anaerobic treatment of wheat starch wastewater from the Candiac, Quebec, plant of Ogilvie Mills Ltd.," Environment Canada, Wastewater Technology Centre, June.

Juby, J.J.G., Leslie, G.L., Deshmukh, S.S., Torres, E.M., Brown, J.P., Sethi, S., and Buhr, H.O. (2000). "A novel membrane-anaerobic digestion approach for wastewater treatment and water reclamation," *2000 Water Reuse Conference Proceedings,* San Antonio, TX, January–February.

Judd, S.J., Le-Clech, P., Taha, T., and Cui, Z.G. (2001). "Theoretical and experimental representation of a submerged MBR system," *Membrane Technology*, No. 135, p. 4.

Kang, I.J. (1996). "Comparison of fouling characteristics between ceramic and polymeric membranes in membrane coupled anaerobic bioreactors," Master's Thesis, Department of Chemical Technology, Seoul National University, Korea.

Kang, I-J., Yoon, S-H., and Lee, C-H. (2002). "Comparison of the filtration characteristics of organic and inorganic membranes in a membrane-coupled anaerobic bioreactor," *Water Research*, Vol. 36, p. 1803.

Kang, I-J., Lee, C-H., and Kim, K-J. (2003). "Characteristics of microfiltration membranes in a membrane coupled sequencing batch reactor system," *Water Research,* Vol. 37, p. 1192.

Kataoka, N., Tokiwa, Y., Tanaka, Y., Fujiki, K., Taroda, H., and Takeda, K. (1992). "Examination of bacterial characteristics of anaerobic membrane bioreactors in three pilot-scale plants for treating low-strength wastewater by application of the colony-forming-curve analysis method," *Applied and Environmental Microbiology,* Vol. 58, No. 9, p. 2751.

Kawayake, E., Narukami, Y., and Yamagata, M. (1991). "Anaerobic digestion by a ceramic membrane enclosed reactor," *Journal of Fermentation and Bioengineering,* Vol. 71, No. 2, p 122.

Kim, J-O. and Somiya, I. (1999). "Performance of organics decomposition and filterability improvement of ceramic membrane by intermittent ozonation in an organic acid fermenter," *Journal of Japanese Sewage Works Association,* Vol. 36, No. 3, p. 101.

Kim, J-O., Somiya, I., Shin, E-B., Bae, W., Kim, S-K., and Kim, R-E. (2002). "Application of membrane-coupled anaerobic volatile fatty acids fermentor for dissolved organics recovery from coagulated raw sludge," *Water Science & Technology,* Vol. 45, No. 12, p. 167.

Kim, J.S., Lee, C.H., and Chang, I.S. (2001). "Effect of pump shear on the performance of a cross-flow membrane bioreactor," *Water Research,* Vol. 36, p. 2137.

Kim, J.S. and Lee, C.H. (2003). "Effect of powdered activated carbon on the performance of an aerobic membrane bioreactor: comparison between cross-flow and submerged membrane systems," *Water Environment Research,* Vol. 75, No. 4, p. 300.

Kimura, K., Nakamura, M., and Watanabe, Y. (2001). "Nitrate removal by a combination of elemental sulfur-based denitrification and membrane filtration," *Water Research,* Vol. 36, p. 1758.

Kiriyama, K., Tanaka, Y., and Mori, I. (1992). "Field test of a composite methane gas production system incorporating a membrane module for municipal sewage," *Water Science & Technology,* Vol. 25, p. 135.

Kirmura, S. (1991). "Japans aqua renaissance '90 project," *Water Science & Technology*, Vol. 23, No. 7-9, p.1573.

Kitamura, Y., Maekawa, T., Tagawa, A., Hayashi, H., and Farrell-Poe, K.L. (1996). "Treatment of strong organic, nitrogenous wastewater by an anaerobic contact process incorporating ultrafiltration," *Applied Engineering in Agriculture,* Vol. 12, No. 6, p 709.

Kobayashi, H.A., Conway De Macario, E., Williams, R.S., and Macario, A.J.L. (1988). "Direct characterization of methanogens in two high-rate anaerobic biological reactors," *Applied and Environmental Microbiology,* Vol. 54, No. 3, p. 693.

Kraft, A. and Brockmann, M. (1999). "Anaerobic technology and membrane bioreactor technology for energy and water generation," *Wasser Abwasser Praxis,* Vol. 8, No. 3, p. 15 (in German).

Lai, L.S., Fakhru'l-Razi, A., Idris, A., and Hassan, M. (1999). "Performance and kinetic study of membrane anaerobic system in treating POME," *Artificial Cells, Blood Substitutes, and Immobilization Biotechnology,* Vol. 27, No. 5, p. 469.

Langenhoff, A.A.M., Intrachandra, N., and Stuckey, D.C. (2000). "Treatment of dilute soluble and colloidal wastewater in an anaerobic baffled reactor: influence of hydraulic retention time," *Water Research,* Vol. 34, No. 4, p. 1307.

Langenhoff, A.A.M. and Stuckey, D.C. (2000). "Treatment of dilute wastewater using an anaerobic baffled reactor: effect of low temperature," *Water Research,* Vol. 34, No. 15, p. 3867.

Lawrence, D., Van Bentem, A., Van Der Herberg and Roeleveld, P. (2001). "MBR pilot research at the Beverwijk WWTP, a step forward to full scale implementation," STOWA, Foundation for Applied Water Research 2001, H_2O, October, p. 11.

Lee, S., Burt, A., Rusoti, G., and Buckland, B. (1999). "Microfiltration of yeast cells using a rotary disk dynamic filtration system," *Biotechnology and Bioengineering*, Vol. 48, p. 386.

Lee, J., Ahn, W-Y., and Lee, C-H. (2001a). "Comparison of the filtration characteristics between attached and suspended growth microorganisms in a submerged membrane bioreactor," *Water Research,* Vol. 35, p. 2435.

Lee, J.C., Kim, J.S., Kang, I.J., Cho, M.H., Park, P.K., and Lee, C.H. (2001b). "Potential and limitations of alum or zeolite addition to improve the performance of a submerged membrane bioreactor," *Water Science & Technology,* Vol. 43, No. 11 p. 59.

Lee, S-M., Jung, J-Y., and Chung, Y-C. (2001c). "Novel method for enhancing permeate flux of submerged membrane system in two-phase anaerobic reactor," *Water Research,* Vol. 35, No. 2, p. 471.

Lei, E. and Berube, P.R. (2004). "Impact of membrane configuration and hydrodynamic conditions on the permeate flux in submerged membrane systems for drinking water treatment," Proceedings British Columbia Water and Waste Association Conference, Whistler, BC, Canada.

Lettinga, G., De Man, A., Van Der Last, A.R.M., Wiegrant, W., Van Knippenberg, K., Frijors, J., and Van Buren, J.C.L. (1993). "Anaerobic treatment of domestic sewage and wastewater," *Water Science & Technology,* Vol. 27, No.9, p. 67.

Lew, B., Belavski, M., Admon, S., Tarre, S., and Green, M. (2003). "Temperature effect on UASB reactor operation for domestic wastewater treatment in temperate climate regions," *Water Science & Technology,* Vol. 48, No. 3, p. 25.

Li, A. and Corrado, J.J. (1986). "Scale-up of the membrane anaerobic reactor system," *Proceedings 40th Industrial Waste Conference,* Purdue University, West Lafayette, Indiana, Ann Arbor Science, Ann Arbor, MI, p. 805.

Lubbecke, A., Vogelpohl, A., and Dewjanin, W. (1995). "Wastewater treatment in a biological high-performance system with high biomass concentration," *WaterResearch*, Vol. 29, p. 793.

Madakoro, T., Ueno, M., Moro, M., Yamamoto, T., and Shibata, T. (1999). "Anaerobic digestion system with micro-filtration membrane for kitchen refuse," Presented at 2nd International Symposium on Anaerobic Digestion of Solid Wastes, Barcelona, Spain, June.

Magara, Y., Nishimura, K., Itoh, M., and Tanaka, M. (1992). "Biological denitrification system with membrane separation for collective human excreta treatment plant," *Water Science & Technology,* Vol. 25, No. 10, p. 241.

Mallevialle, J., Odendaal, P.E., and Wiesner, M.R. (editors) (1996). "Water treatment membrane processes," McGraw-Hill Publishing, New York, NY, USA.

Mamais, D., Paul, A.P., Cheng, Y.W., Loiacono, J., and Jenkins, D. (1994). "Determination of ferric chloride dose to control struvite precipitation in anaerobic sludge digesters," *Water Environment Research,* Vol. 66, No. 7, p. 912.

Manning, J., Sutton, P.M., Gaines, F.R., and Dunn, W.G. (2001). "Combined and separate municipal and leachate wastewater treatment: full scale application of the membrane bioreactor system," Proceedings of the WEF 74th Annual Conference & Exposition, Atlanta, GA, CD-ROM, October.

Minami, K., Okamura, K., Ogawa, S., and Naritomi, T. (1991). "Continuous anaerobic treatment of wastewater from a kraft pulp mill," *Journal of Fermentation and Bioengineering*, Vol. 71, No. 4, p. 270.

Mizuno, O., Takagi, H., and Noike, T. (1998). "Biological sulphate removal in an acidogenic bioreactor with an ultrafiltration membrane system," *Water Science & Technology*, Vol. 38, No. 4-5, p. 513.

Morgenroth, E., Raskin, L., Norddahl, B., Criddle, C., and Hickey, B. (2003). "Membrane bioreactors for biological wastewater treatment and water reuse –advanced treatment of swine manure and how it relates to CAMPWS," CAMPWS Seminar Presentation, University of Illinois at Urbana-Champaign, November.

Munariatis, I.D. and Grigoropoulos, S.G. (2002). "Low-strength wastewater treatment using an anaerobic baffled reactor," *Water Environment Research*, Vol. 74, No., 2, p.170.

Nagano, A., Arikawa, E., and Kobayashi, H. (1992). "The treatment of liquor wastewater containing high-strength suspended solids by membrane bioreactor system," *Water Science & Technology,* Vol. 26, No. 3-4, P. 887.

Nagaoka, H. (1999). "Nitrogen removal by submerged membrane separation activated sludge process," *Water Science & Technology,* Vol. 39, No. 8, p. 107.

Ng, W.J., Ong, S.L., Gomez, M.J., Hu, J.Y., and Fan, X.J. (2000). "Study on a sequencing batch membrane bioreactor for wastewater treatment," *Water Science & Technology,* Vol. 41, No. 10-11, p. 227.

Nikaido, S., Nishimura, K., and Noike, T. (1996). "Study on treatment of long-chain fatty acid containing wastewater in the anaerobic digestion with membrane filtration systems," *J. Japan Society on Water Environment,* Vol. 19, No. 3, p. 220 (in Japanese).

Norddahl, B. and Rohold, L. (1998). "Biorek principle - a membrane bioreactor coupled to an ammonia stripper and an RO unit enabling conversion of manure and slurry into energy, concentrated fertiliser and potable water," Proceedings of Bioenergy 98, Madison, WI, October.

Nuhoglu, A., Pekdemir, T., Yildiz, E., Keskinler, B., and Akay, G. (2002). "Drinking water denitrification by a membrane bio-reactor," *WaterResearch,* Vol. 36, p. 1155.

Ognier, S., Wisniewski, C., and Grasmick, A. (2001). "Biofouling in membrane bioreactors: phenomenon analysis and modeling," Presented at Membrane Bioreactors — International Conference, May.

Okamura, K. (1994). "Study of wastewater treatment using anaerobic bioreactor combined UF membrane," *Papers — Ship Research Institute Tokyo*, Vol. 31, No. 5, p. 2–50.

Omstead, D.R., Jeffries, T.W., Naughton, R., and Gregor, H.P. (1980). "Membrane controlled digestion: anaerobic production of methane and organic acids," *Biotechnology & Bioengineering Symposium, No. 10,* New York: John Wiley & Sons, p. 247.

Onysko, K.A. and Hall, E.R. (1993). "Anaerobic membrane bioreactor treatment of segregated bleach plant wastewater." 1993 Environmental Conference, Tappi, p. 845.

O'Sullivan, T.J., Epstein, A.C., Korchin, S.R., and Beaton, N.C. (1984). "Applications of ultrafiltration in biotechnology," *Chemical Engineering Progress*, January, p. 68.

Ozaki, N. and Yamamoto, K. (2001). "Hydraulic effects on sludge accumulation on membrane surface in crossflow filtration," *Water Research*, Vol. 35, No. 13, p. 3137.

Pagilla, K.R. (2004). Personal communication to P.M. Sutton, Associate Professor, Illinois Institute of Technology, February.

Park, H., Choo, K., and Lee, C. (1999). "Flux enhancement with powdered activated carbon addition in the membrane anaerobic bioreactor," *Separation Science and Technology,* Vol. 34, No. 14, p. 2781.

Pillay, V.L., Townsend, B., and Buckley, C.A. (1994). "Improving the performance of anaerobic digesters at waste water treatment works: the coupled cross-flow microfiltration/digester process," *Water Science & Technology*, Vol. 30, No. 12, p. 329.

Pirbazari, M., Ravindran, V., Badriyha, B.N., and Kim, S.H. (1996). "Hybrid membrane filtration process for leachate treatment," *Water Research,* Vol. 30, No. 11, p. 2691.

Rebac, S. (1998). "Psychrophilic anaerobic treatment of low strength wastewaters," Wageningen Agricultural University, Wageningen, The Netherlands.

Ross, B. and Strohwald, H. (1994). "Membranes add edge to old technology," *Water Quality International,* No. 4, , p 18.

Ross, W.R., Barnard, J.P., Le Roux, J., and Villiers, H.A. (1990). "Application of ultrafiltration membranes for solids-liquid separation in anaerobic digestion systems: The ADUF process," *Water Science & Technology*, Vol. 16, p. 85.

Ross, W.R., Barnard, J.P., Strohwald, N.K.H., Grobler, C.J., and Sanetra, J. (1992). "Practical application of the ADUF process to the full-scale treatment of a maize-processing effluent," *Water Science & Technology*, Vol. 25, No. 10, p 27.

Ross W.R., Strohwald, N.K.H., Grobler, C.J., and Senetra, J. (1994). "Membrane-assisted anaerobic treatment of industrial effluents: The South African ADUF process," *Proceedings of the 7th International Symposium on Anaerobic Digestion,* Cape Town, South Africa, p. 550.

Sainbayar, A., Kim, J.S., Jung, W.J., Lee, Y.S., and Lee, C.H. (2001). "Application of surface modified polypropylene membranes to an anaerobic membrane bioreactor," *Environmental Technology,* Vol. 22, No. 9, p. 1035.

Saw, C.B., Anderson, G.K., James, A., and Le, M.S. (1986). "A membrane technique for biomass retention in anaerobic waste treatment," *Proceedings 40th Industrial Waste Conference,* Purdue University, West Lafayette, IN. Ann Arbor, MI: Ann Arbor Science, p. 805.

Schiener, P., Nachaiyasit, S., and Stuckey, D.C. (1998). "Production of soluble microbial products (SMP) in an anaerobic baffled reactor: composition biodegradability and the effect of process parameters," *Environmental Technology,* Vol. 19, No. 4, p. 391.

Seghezzo, L., Guerra, R.G., Gonzalez, S.M., Trupiano, A.P., Figueroa, M.E., Cuevas, C.M., Zeeman, G., and Lettinga, G. (2002). "Removal efficiency and methanogenic activity profiles in a pilot-scale UASB reactor treating settled sewage at moderate temperatures," *Water Science & Technology,* Vol. 45, No. 10, p. 243.

Shimizu, U., Rokudai, M., Tohya, S., Kauawake, E., Yazawa, T., Tanaka, H., and Eguchi, K. (1989). "Effect of pore size on the filtration characteristics of ceramic membrane for membrane bioreactors," *Kagaku Kogaku Ronbunshu,* Vol. 15, p. 322.

Shimizu, Y., et al. (1991). "Influence of organic acid density in methane fermentation liquid on membrane filtration characteristics," *Japanese Journal of Fermentation Technology,* Vol. 67, No. 1, p. 15.

Shimizu, Y., Matsushita, K., and Watanabe, A. (1994). "Influence of shear breakage of microbial cells on cross-flow microfiltration flux," *Journal of Fermentation and Bioengineering,* Vol. 78, No. 2, p. 170.

Shimizu, Y., Uryu, K., Okuno, Y., and Watanabe, A. (1996). "Cross-flow microfiltration of activated sludge using submerged membrane with air bubbling," *Journal of Fermentation and Bioengineering,* Vol. 81, No. 1, p. 55.

Shimizu, Y., Okuno, Y., Uryu, K., Ohtsubo, S., and Watanabe, A. (1996). "Filtration characteristics of hollow fiber microfiltration membranes used in membrane bioreactor for domestic wastewater treatment," *Water Research,* Vol. 30, No. 10, p. 2385.

Shin, H. and Kang, S. (2002). "Performance and membrane fouling in a pilot scale SBR process coupled with membrane," *Water Science & Technology,* Vol. 47, No. 1, p. 139.

Singh, K.S. and Viraraghavan, T. (2003). "Impact of temperature on performance, microbiological, and hydrodynamic aspects of UASB reactors treating municipal wastewater," *Water Science & Technology,* Vol. 48, No. 6, p. 211.

Smith, C.V., Di Gregorio, D.O., and Talcott, R.M. (1969). "The use of ultrafiltration membranes for activated sludge separation," Proceedings 24th Industrial Waste Conference, Purdue University, West Lafayette, Indiana, p. 805.

Stensel, H.D. and Strand, S.E. (2004). "Evaluation of feasibility of methods to minimize biomass production from biotreatment," Water Environment Research Foundation, Project Report 00-CTS-10 (in publication).

Strohwald, N.K.H. and Ross, W.R. (1992). "Application of the ADUF process to brewery effluent on a laboratory scale," *Water Science & Technology,* Vol. 25, No. 10, p 95.

Stuckey, D.C. (2003). "The submerged anaerobic membrane bioreactor (SAMBR): an intensification of anaerobic wastewater treatment," Abstract of Presentation to the Department of Civil Engineering at the University of Minnesota, Minneapolis, MN, September.

Stuckey, D.C. and Hu, A. (2003). "The submerged anaerobic membrane bioreactor (SAMBR): an intensification of anaerobic wastewater treatment," Presented at the International Water Association Leading Edge Conference on Drinking Water and Wastewater Treatment Technologies, Noordwijk, The Netherlands, May.

Sutton, P.M., Li, A., Evans R.R., and Korchin S.R. (1983). "Dorr-Oliver's fixed film and suspended growth anaerobic systems for industrial wastewater treatment and energy recovery," Proceedings 37th Industrial Waste Conference, Purdue University, West Lafayette, IA. Ann Arbor, MI: Ann Arbor Science, p. 667.

Sutton, P.M. and Evans, R.R. (1983). "Anaerobic system designs for efficient treatment of industrial wastewater," Proceedings of the 3rd Symposium on Anaerobic Digestion, IAWQ.

Sutton, P.M. (1986a). "Active biomass retention: the key to anaerobic process efficiency," Presented at Anaerobic Fixed-Film Digestion Seminar, Ontario Ministry of Health Laboratories, Toronto, Ontario, November.

Sutton, P.M. (1986b). "Innovative biological systems for anaerobic treatment of grain and food processing wastewater," *Starch/Starke,* Vol. 38, No. 9, p.314.

Sutton, P.M., Mishra, P.N., Bratby, J.R., and Enegess, D. (2002). "Membrane bioreactor industrial and municipal wastewater applications: long-term operating experience," Presented at the WEF 75th Annual Conference & Exposition, Chicago, IL, September–October.

Takashima, M., Kudoh, Y., and Tabata, N. (1996). "Complete anaerobic digestion of activated sludge by combining membrane separation and alkaline heat post-treatment," *Water Science & Technology*, Vol. 34, No. 5–6, p. 3.

Tardieu, E., Grasmick, A., Geaugey, V., and Manem, J. (1999). "Influence of hydrodynamics on fouling velocity in a recirculated MBR for wastewater treatment," *Journal of Membrane Science*, Vol. 156, No. 10, p. 131.

Torres, P. and Foresti, E. (2001). "Domestic sewage treatment in a pilot system composed of UASB and SBR reactors," *Water Science & Technology*, Vol. 44, No. 4, p. 247.

Uemura, S. and Harada, H. (2000). "Treatment of sewage by a UASB reactor under moderate to low temperature conditions," *Bioresource Technology*," Vol. 72. No. 3, p. 275.

Uemura, T. and Kondou, Y. (2003). "Membrane technology progress in Japan," *Water 21*, August.

Van Bentem, A., Lawrence, D., Horjus, F., De Boer, R., Koornneef. E., and Van Efferen, B. (2001). "MBR pilot research in Beverwijk: side studies," STOWA, Foundation for Applied Water Research 2001, H_2O, October, p. 16.

Van Der Last, A.R.M., and Lettinga, G. (1992). "Anaerobic treatment of domestic wastewater under moderate climatic (Dutch) conditions using upflow reactors at increased superficial velocities," *Water Science & Technology*, Vol. 25, No. 7, p. 167.

Van Der Roest, H. and Leenen, J. (2001). "The MBR as worldwide promising technology," STOWA, Foundation for Applied Water Research 2001, H_2O, October, p. 5.

Van Houten, R., Evenblij, H., and Keijmel, M. (2001). "Membrane bioreactors hit the big time – ten years of research in the Netherlands," STOWA, Foundation for Applied Water Research 2001, H_2O, October, p. 26.

Vera, L., Delgado, S., and Elmaleh, S. (2000). "Gas-sparged cross-flow microfiltration of biologically treated wastewater," *Water Science & Technology*, Vol. 41, No. 10-11, p. 173.

Weigant, W.M. (2001). "Experiences and potential of anaerobic wastewater treatment in tropical regions," *Water Science & Technology*, Vol. 44, No. 8, p. 107.

Wen, W., Huang, X., and Qian, Y. (1999). "Domestic wastewater treatment using an anaerobic bioreactor coupled with membrane filtration," *Process Biochemistry*, Vol. 35, No. 3–4, p. 335.

Wisniewski, C. and Grasmick, A. (1998). "Floc size distribution in a membrane bioreactor and consequences for membrane fouling," *Colloids and Surfaces, A: Physicochemical and Engineering Aspects*, Vol. 138, p. 403.

Yamagishi, S., et al. (1997). "Study on the factors influencing permeate flux of membrane coupled with anaerobic membrane reactor," *Japanese Industrial Water*, Vol. 468, p. 2.

Yamagishi, T., Matsuda, M., Masuda, H., Kamisawa, C., Iizumi, S., Oomura, T., Nagura, K., Ito, S., Taked, A.H., Suzuki, T., Toyohara, H., and Okada, T. (1990). "Cross-flow filtration for anaerobic wastewaters," *Polution Japan*, Vol. 5, p. 291.

Yamaguchi, M., Okamura, K., Tanimoto, Y., Naritomi, T., Ogawa, S., Hake, J., and Minami, K. (1990). "Thermophilic methane fermentation of evaporator condensate from a kraft pulp mill," TAPPI Environmental Conference, Seattle, WA, Vol. 2, p. 631.

Yamamoto, K., Hiasa, M., Mahmood, T., and Matsuo, T. (1989). "Direct solid-liquid separation using hollow fiber membrane in an activated sludge aeration tank," *Water Science & Technology,* Vol. 21, p. 43.

Yiliang, H., Zhichao, W., Chunjie, L., Guowie, G., Fenging, X., Mingxu, Z., and Zuyi, C. (1999). "Application of the anaerobic MBR for treatment of high concentration food wastewater," *Chinese Journal of Environmental Science,* November (in Chinese).

Yoon, S.H., Kang, I.J., and Lee, C.H. (1999). "Fouling of inorganic membrane and flux enhancement in membrane-coupled anaerobic bioreactor." *Separation Science and Technology,* Vol. 34, No. 5, p.709.

Yuntao, G., Zhanpeng, J., Wanpeng, Z., and Zhongying, C. (1998). "Two-phase anaerobic membrane biosystem for organic wastewater treatment," *Chinese Journal of Environmental Science,* Vol. 21, No. 4, p. 52 (in Chinese).

Yuntao, G., Zhanpeng, J., Wanpeng, Z., and Peng, J. (2000). "Two-phase anaerobic membrane biosystem for treatment of papermill wastewater," *Chinese Journal of Environmental Science,* Vol. 19, No. 6, p. 56 (in Chinese).

Yushina, Y. and Hasegawa, J. (1994). "Process performance comparison of membrane introduced anaerobic digestion using food industry waste water," *Desalination*, Vol. 98, No. 1–3, p 413.

Zeeman, G. and Lettinga, G. (2002). "The role of anaerobic digestion of domestic sewage in closing the water and nutrient cycle at community level," *Water Science & Technology*, Vol. 39, No. 5, p. 187.

Zhang, Z. and Hall, E.R. (2006). "Heterotrophic kinetic parameter estimation for enhanced biological phosphorus removal processes operated in conventional and membrane assisted modes," *Water Quality Research Journal of Canada,* Vol. 41, p. 72.

Zoh, K-D. and Stenstrom, M.K. (2001). "Application of a membrane bioreactor for treating explosives process wastewater," *Water Research,* Vol. 36, p. 1,018.

WASTEWATER UTILITY

Alabama
Montgomery Water Works & Sanitary Sewer Board

Alaska
Anchorage Water & Wastewater Utility

Arizona
Gila Resources
Glendale, City of, Utilities Department
Mesa, City of
Peoria, City of
Phoenix Water Services Department
Pima County Wastewater Management

Arkansas
Little Rock Wastewater Utility

California
Calaveras County Water District
Central Contra Costa Sanitary District
Corona, City of
Crestline Sanitation District
Delta Diablo Sanitation District
Dublin San Ramon Services District
East Bay Dischargers Authority
East Bay Municipal Utility District
Eastern Municipal Water District
El Dorado Irrigation District
Fairfield-Suisun Sewer District
Fresno Department of Public Utilities
Inland Empire Utilities Agency
Irvine Ranch Water District
Las Virgenes Municipal Water District
Livermore, City of
Lodi, City of
Los Angeles, City of
Napa Sanitation District
Orange County Sanitation District
Palo Alto, City of
Riverside, City of
Sacramento Regional County Sanitation District
San Diego Metropolitan Wastewater Department, City of
San Francisco, City & County of
Sanitation Districts of Los Angeles County
San Jose, City of
Santa Barbara, City of
Santa Cruz, City of
Santa Rosa, City of
South Bayside System Authority
South Coast Water District
South Orange County Wastewater Authority
Stege Sanitary District
Sunnyvale, City of
Union Sanitary District
West Valley Sanitation District

Colorado
Aurora, City of
Boulder, City of
Colorado Springs Utilities
Greeley, City of
Littleton/Englewood Water Pollution Control Plant
Metro Wastewater Reclamation District, Denver

Connecticut
The Mattabassett District
Greater New Haven WPCA

District of Columbia
District of Columbia Water & Sewer Authority

Florida
Broward, County of
Fort Lauderdale, City of
JEA
Miami-Dade Water & Sewer Authority
Orange County Utilities Department
Reedy Creek Improvement District
Seminole County Environmental Services
St. Petersburg, City of
Stuart Public Utilities
Tallahassee, City of
Tampa, City of
Toho Water Authority
West Palm Beach, City of

Georgia
Atlanta Department of Watershed Management
Augusta, City of
Clayton County Water Authority
Cobb County Water System
Columbus Water Works
Fulton County
Gwinnett County Department of Public Utilities
Savannah, City of

Hawaii
Honolulu, City & County of

Idaho
Boise, City of

Illinois
American Bottoms Wastewater Treatment Plant
Greater Peoria Sanitary District
Kankakee River Metropolitan Agency
Metropolitan Water Reclamation District of Greater Chicago
Wheaton Sanitary District

Iowa
Ames, City of
Cedar Rapids Wastewater Facility
Des Moines, City of
Iowa City

Kansas
Johnson County Unified Wastewater Districts
Unified Government of Wyandotte County/ Kansas City, City of

Kentucky
Louisville & Jefferson County Metropolitan Sewer District

Louisiana
Sewerage & Water Board of New Orleans

Maine
Bangor, City of
Portland Water District

Maryland
Anne Arundel County Bureau of Utility Operations
Hagerstown, City of
Howard County Department of Public Works
Washington Suburban Sanitary Commission

Massachusetts
Boston Water & Sewer Commission
Upper Blackstone Water Pollution Abatement District

Michigan
Ann Arbor, City of
Detroit, City of
Holland Board of Public Works
Lansing, City of
Saginaw, City of
Wayne County Department of Environment
Wyoming, City of

Minnesota
Rochester, City of
Western Lake Superior Sanitary District

Missouri
Independence, City of
Kansas City Missouri Water Services Department
Little Blue Valley Sewer District
Metropolitan St. Louis Sewer District

Nebraska
Lincoln Wastewater System

Nevada
Henderson, City of
Reno, City of

New Jersey
Bergen County Utilities Authority
Ocean, County of
Passaic Valley Sewerage Commissioners

New York
New York City Department of Environmental Protection

North Carolina
Charlotte/Mecklenburg Utilities
Durham, City of
Metropolitan Sewerage District of Buncombe County
Orange Water & Sewer Authority

Ohio
Akron, City of
Butler County Department of Environmental Services
Columbus, City of
Metropolitan Sewer District of Greater Cincinnati
Northeast Ohio Regional Sewer District
Summit, County of

Oklahoma
Tulsa, City of

Oregon
Clean Water Services
Eugene/Springfield Water Pollution Control
Water Environment Services

Pennsylvania
Philadelphia, City of
University Area Joint Authority

South Carolina
Charleston Commissioners of Public Works
Mount Pleasant Waterworks & Sewer Commission
Spartanburg Sanitary Sewer District

Tennessee
Cleveland, City of
Knoxville Utilities Board
Murfreesboro Water & Sewer Department
Nashville Metro Water Services

Texas
Austin, City of
Dallas Water Utilities
Denton, City of
El Paso Water Utilities
Fort Worth, City of
Gulf Coast Waste Disposal Authority
Houston, City of
San Antonio Water System
Trinity River Authority

Utah
Salt Lake City Corporation

Virginia
Alexandria Sanitation Authority

WERF SUBSCRIBERS

Arlington, County of
Fairfax County
Hampton Roads Sanitation District
Henrico, County of
Hopewell Regional Wastewater Treatment Facility
Loudoun County Sanitation Authority
Lynchburg Regional WWTP
Prince William County Service Authority
Richmond, City of
Rivanna Water & Sewer Authority

Washington
Everett, City of
King County Department of Natural Resources
Seattle Public Utilities
Sunnyside, Port of
Yakima, City of

Wisconsin
Green Bay Metro Sewerage District
Kenosha Water Utility
Madison Metropolitan Sewerage District
Milwaukee Metropolitan Sewerage District
Racine, City of
Sheboygan Regional Wastewater Treatment
Wausau Water Works

Australia
South Australian Water Corporation
Sydney Water Corporation
Water Corporation of Western Australia

Canada
Greater Vancouver Regional District
Regina, City of, Saskatchewan
Toronto, City of, Ontario
Winnipeg, City of, Manitoba

New Zealand
Watercare Services Limited

United Kingdom
Yorkshire Water Services Limited

STORMWATER UTILITY

California
Los Angeles, City of, Department of Public Works
Monterey, City of
San Francisco, City & County of
Santa Rosa, City of

Colorado
Aurora, City of
Boulder, City of

Georgia
Griffin, City of

Iowa
Cedar Rapids Wastewater Facility
Des Moines, City of

Kansas
Overland Park, City of

Maine
Portland Water District

Minnesota
Western Lake Superior Sanitary District

North Carolina
Charlotte, City of, Stormwater Services

Pennsylvania
Philadelphia, City of

Tennessee
Chattanooga Stormwater Management

Washington
Bellevue Utilities Department
Seattle Public Utilities

STATE

Arkansas Department of Environmental Quality
Fresno Metropolitan Flood Control District
Kansas Department of Health & Environment
Kentucky Department of Environmental Protection
Ohio River Valley Sanitation Commission
Urban Drainage & Flood Control District, CO

CORPORATE

Abt. Associates
ADS Environmental Services
Alan Plummer & Associates
Alden Research Laboratory Inc.
American Water
Aqua-Aerobic Systems Inc.
Aquateam–Norwegian Water Technology Centre A/S
ARCADIS
Associated Engineering
Black & Veatch
Blasland, Bouck & Lee Inc.
Boyle Engineering Corporation
Brown & Caldwell
Burns & McDonnell
CABE Associates Inc.
The Cadmus Group
Camp Dresser & McKee Inc.
Carollo Engineers Inc.
Carpenter Environmental Associates Inc.
CDS Technologies Inc.
Chemtrac Systems Inc.
CH2M HILL
Construction & Environmental Consulting, Inc.
D&B/Guarino Engineers, LLC
Damon S. Williams Associates, LLC
Dewling Associates Inc.
Earth Tech Inc.
Ecovation
EMA Inc.
Environ/The ADVENT Group, Inc.
The Eshelman Company Inc.
EWT Holdings Corporation
Fay, Spofford, & Thorndike Inc.
Freese & Nichols Inc.
ftn Associates Inc.
Fuss & O'Neill Inc.
Gannett Fleming Inc.
Geosyntec Consultants
GHD
Golder Associates Ltd.
Greeley and Hansen LLC
Hazen & Sawyer, P.C.
HDR Engineering Inc.
HNTB Corporation
HydroQual Inc.
Infilco Degremont Inc.
Jacobson Helgoth Consultants Inc.
Jason Consultants LLC Inc.
Jordan, Jones, & Goulding Inc.
KCI Technologies Inc.
Kelly & Weaver, P.C.
Kennedy/Jenks Consultants
KMK Consultants
Komline Sanderson Engineering Corporation
Lawler, Matusky & Skelly Engineers, LLP
Limno-Tech Inc.
Lombardo Associates Inc.
Malcolm Pirnie Inc.
Material Matters
McKim & Creed
MEC Analytical Systems Inc.
Metcalf & Eddy Inc.
MPR Engineering Corporation, Inc.
MWH
New England Organics
O'Brien & Gere Engineers Inc.
Odor & Corrosion Technology Consultants Inc.
Oswald Green, LLC
PA Government Services Inc.
Parametrix Inc.
Parsons
Post, Buckley, Schuh & Jernigan
R&D Engineering/Conestoga Rover & Associates
RMC Inc.
R.M. Towill Corporation
Ross & Associates Ltd.
Rothberg, Tamburini & Windsor, Inc.
Royce Technologies
Stantec Consulting Inc.
Stearns & Wheler, LLC
Stone Environmental Inc.
Stormwater360
Stratus Consulting Inc.
Synagro Technologies Inc.
Tetra Tech Inc.
Trojan Technologies Inc.
Trussell Technologies, Inc.
URS Corporation
USfilter
Wade-Trim Inc.
Westin Engineering Inc.
Weston Solutions Inc.
Woodard & Curran
WRc/D&B, LLC
WWETCO, LLC
Zenon Environmental Inc.
Zoeller Pump Company

INDUSTRY

American Electric Power
ChevronTexaco Energy Research & Technology Company
The Coca-Cola Company
Dow Chemical Company
DuPont Company
Eastman Chemical Company
Eastman Kodak Company
Eli Lilly & Company
Merck & Company Inc.
Premier Chemicals LLC
Procter & Gamble Company
RWE Thames Water Plc
Severn Trent Services Inc.
Suez Environnment
United Water Services LLC

Note: List as of 4/1/06

WERF Board of Directors

WERF Research Council

WERF Product Order Form

a benefit of joining the Water Environment Research Foundation, subscribers are entit ed to receive one complimentary copy of all final orts and other products. Additional copies are available at cost (usually $10). To order your complimentary copy of a report, please write e" in the unit price column. WERF keeps track of all orders. If the charge differs from what is shown here, we will call to confirm the total ore processing.

ne | Title

anization

lress

| State | Zip Code | Country

one | Fax | Email

Stock #	Product	Quantity	Unit Price	Total
			Postage & Handling	
			VA Residents Add 4.5% Sales Tax	
			Canadian Residents Add 7% GST	
			TOTAL	

Method of Payment: (All orders must be prepaid.)

Check or Money Order Enclosed

Visa ❑ Mastercard ❑ American Express

ount No. | Exp. Date

nature

Shipping & Handling:

Amount of Order	United States	Canada & Mexico	All Others
Up to but not more than:	Add:	Add:	Add:
$20.00	$5.00*	$8.00	50% of amount
30.00	5.50	8.00	40% of amount
40.00	6.00	8.00	
50.00	6.50	14.00	
60.00	7.00	14.00	
80.00	8.00	14.00	
100.00	10.00	21.00	
150.00	12.50	28.00	
200.00	15.00	35.00	
More than $200.00	Add 20% of order	Add 20% of order	

*minimum amount for all orders

te: Please make checks payable to the Water Environment Research Foundation.

To Order (Subscribers Only):

Log on to www.werf.org and click on the "Product Catalog."

Phone: (703) 684-2470
Fax: (703) 299-0742.

WERF
Attn: Subscriber Services
635 Slaters Lane
Alexandria, VA 22314-1177

To Order (Non-Subscribers):

Non-subscribers may be able to order WERF publications either through WEF (www.wef.org) or IWAP (www.iwapublishing.com). Visit WERF's website at www.werf.org for details.

IWA Publishing's authorised EU representative for General Product Safety Regulations is Diane D'Arras, 15 rue Duret, 75116 Paris, France, e-mail: safety@iwap.co.uk.

Printed and bound by CPI Group (UK) Ltd, Croydon, CR0 4YY
12/05/2026
02108874-0010